THE PHYSICS OF PULSARS

Edited by
ALLEN M. LENCHEK
Department of Physics
University of Maryland
College Park, Maryland

GORDON AND BREACH, SCIENCE PUBLISHERS
New York • London • Paris

Editorial office for the United States of America
Gordon and Breach, Science Publishers, Inc.
440 Park Avenue South
New York, N.Y. 10016

Editorial office for France
Gordon & Breach
7–9 rue Emile Dubois
Paris 14e

Library of Congress catalog card number 76-150793. ISBN *cloth:* 0 677 14290 0; *paper:* 0 677 14295 1.

Printed in the United States of America

FOREWORD

This volume contains a series of lectures that were given at the University of Maryland, College Park, in the fall of 1969, including material contributed as late as October, 1970. The series was part of a continuing collaboration with the Goddard Space Flight Center, Greenbelt, Maryland, in presenting lectures on astrophysics at the University of Maryland. It is a pleasure to acknowledge the inestimable contribution made by Dr. Stephen P. Maran of the Goddard Space Flight Center in helping to organize this series.

Most of the articles were transcribed from the lecture by the Editor and then revised by the Author. Chapters 4, 10, 11 and 12 were written entirely by their authors following the lecture. The Appendix contains periods, equatorial coordinates and dispersion measures for the 55 pulsars announced as of September 1970.

My thanks go to Mrs. Ruth Lates who typed most of the book and to the able secretary of the space physics group, Mrs. Rose Pryor, for her capable assistance with the logistics of the lecture series as well as her contribution to the production of the book. To the authors of the several chapters, who provided the Department of Physics and Astronomy with several months of high intellectual excitement, I am deeply indebted.

Allen M. Lenchek

College Park, Maryland

List of Authors

Hale V. Bradt	Center for Space Research, Massachusetts Institute of Technology, Cambridge, Massachusetts.
Hong-Yee Chiu	Institute for Space Studies, Goddard Space Flight Center, New York, New York.
George S. Downs	Jet Propulsion Laboratory, Pasadena, California.
Thomas Gold	Department of Astronomy, Cornell University, Ithaca, New York.
Peter Goldreich	Planetary Sciences Department, California Institute of Technology, Pasadena, California.
Robert M. Hjellming	National Radio Astronomy Observatory, Green Bank, West Virginia.
G. Richard Huguenin	Department of Physics and Astronomy, University of Massachusetts, Amherst, Massachusetts.
Stephen P. Maran	Advanced Systems and Ground Observations Branch, Laboratory for Solar Physics, Goddard Space Flight Center, Greenbelt, Maryland.
S. B. Modali	Laboratory for Solar Physics, Goddard Space Flight Center, Greenbelt, Maryland.
J. P. Ostriker	Princeton University Observatory, Princeton, New Jersey.
F. Pacini	Center for Radiophysics and Space Research, Cornell University, Ithaca, New York.
David H. Staelin	Massachusetts Institute of Technology, Cambridge, Massachusetts.
Yervant Terzian	Center for Radiophysics and Space Research, Cornell-Sydney University Astronomy Center, Cornell University, Ithaca, New York.
E. J. Wampler	Lick Observatory, University of California, Santa Cruz, California.

TABLE OF CONTENTS

PROBLEMS AND PLEASURES OF PULSARS

Stephen P. Maran

Advanced Systems and Ground Observations Branch
Laboratory for Solar Physics
Goddard Space Flight Center

From ancient Chinese and Japanese records we know that in July, 1054 A.D. an exceedingly bright star appeared in the heavens. Visible in daylight for 3 weeks and at night for $\sim$ 600 nights, and also recorded in Arizona canyon and cave paintings, this supernova is now identified with the Crab Nebula, site of the pulsar NP 0532.

The Crab Nebula shines with a light attributed to synchrotron radiation emitted by relativistic electrons, and this radiation extends over the electromagnetic spectrum from the radio region to the hard X-rays. A long-standing puzzle about the Crab is the fact that electrons emitting the X-ray synchrotron radiation should have half-lives of the order of a year, much shorter than the age of the nebula (915 years). Therefore we need to explain the continuous production of energetic electrons in the Crab. The discovery of a pulsar in the Crab now seems to provide the key to solving this problem.

As of 1 September, 37 pulsars had been announced. Many of them are located in the Milky Way, and the majority were discovered by the Australian astronomers at Molonglo. Pulsars were first discovered by the Cambridge group (Hewish, Bell, Pilkington, Scott and Collins, 1968) who had built a wire dipole array designed to observe rapid fluctuations for the study of interplanetary scintillations.

Figure 1 shows the optical pulses from the Crab Nebula pulsar (NP 0532). There is a main pulse followed by an interpulse. In the optical region, the pulse shape is constant with wavelength, but in the radio region the shape is wavelength-dependent. For example, in the Crab pulsar, the main pulse is split when observed at 300 MHz and the interpulse practically disappears at 60 MHz.

A theory of pulsars must explain three points: (1) the clock mechanism, (2) the radiation mechanism, and (3) the beaming mechanism. All current theories are based on the assumption that the clock mechanism is provided by a rotating neutron star. Neutron stars have been studied theoretically for several decades without any evidence for their existence until the discovery of the pulsars. Produced by gravitational collapse during a supernova event, a neutron star is a condensed object with a radius about

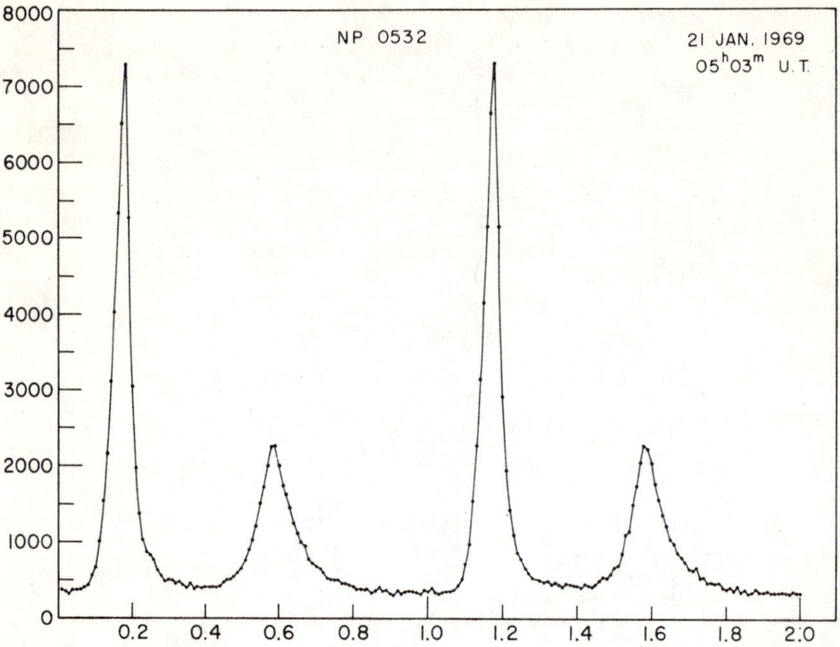

Fig. 1 - Optical pulses from the Crab Nebula pulsar, NP 0532, obtained by R. Lynds, S.P. Maran and D.E. Trumbo with the 84-inch telescope on Kitt Peak.

the distance from College Park to the Goddard Space Flight Center (10 km) and a central density comparable to that of nuclear matter ($\lesssim 10^{15}$ g cm^{-3}). In a classic paper, Wheeler (1966) concluded that it was unlikely that sufficient energy could be stored in vibrational modes of a neutron star to explain the present emissions of the Crab Nebula and he suggested that the possibility of rotational energy storage deserved consideration. In the collapse, the initial magnetic field of the star is compressed to a very large intensity, of the order of 10^{10} to 10^{12} gauss in the neutron star configuration. The possibility that a rotating, magnetized neutron star might power the Crab was studied by Pacini (1967) before the discovery of pulsars but his paper went generally unnoticed.

Two schools of thought exist on the question of just where in the pulsar the pulsed radiation is generated. According to one class of theories, the radiation originates on the surface of the neutron star in the vicinity of the magnetic poles (cf. Chiu and Occhionero 1969), as in Figure 2. The other view generally has the radiation arising at the velocity of light cylinder. This cylinder has a radius r, such that $\Omega r = c$ where c is the speed of light and the angular velocity $\Omega = 2\pi/P$, P being the pulsar period. The pulsars are slowing down, and several braking mechanisms have been suggested. These include loss of angular momentum in a stellar wind (Michel and Tucker, 1969), magnetic dipole radiation (Pacini 1968, Gunn and Ostriker 1969), and gravitational quadropole radiation (Ostriker and Gunn 1969).

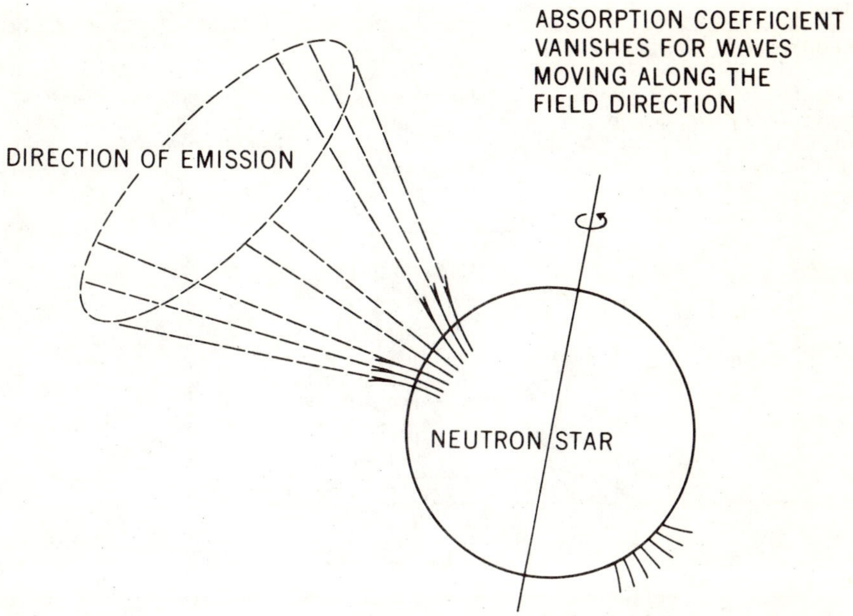

Fig. 2 - Artist's conception of the geometry involved in the Intense Magnetic Field theory of pulsars, as developed by H.Y. Chiu, V. Canuto, L. Fassio-Canuto and F. Occhionero.

A significant feature of the pulsar radiation is its polarization. The radiation is typically linearly polarized; in a few pulsars the plane of polarization makes a smooth, reproducible rotation in position angle during the pulse (cf. Radhakrishnan and Cooke 1969). The presence of linear polarization is in accord with a synchrotron process and the rotation of the plane of polarization within the pulse is consistent with the varying aspect of the source as it is carried around by the rotation. The difference in position angle at a given phase of a pulse when observed at different frequencies was studied in the Vela pulsar (PSR 0833-45) and found to be wholly due to interstellar Faraday rotation.

An important effect is the interstellar dispersion. Free interstellar electrons affect the propagation of the radiation, slowing down the lower frequency waves. The difference in arrival time of pulses of differing frequency, related to the interstellar electron density by

$$\frac{d\nu}{dt} = - \frac{c\nu^3}{\int \nu_p^{\,2}\, ds}, \qquad (1)$$

where s is the distance to the pulsar, $\nu_p \propto \sqrt{N_e}$ is the plasma frequency and N_e is the electron density, leads to a determination of DM, the dispersion measure, given by

$$DM = \int N_e \, ds. \tag{2}$$

Dispersion measures of pulsars show a wide range of values for pulsars that lie close to the galactic plane but they are relatively small for pulsars at high galactic latitude (Figure 3). This distribution is characteristic of objects that are generally in or near the galactic disc.

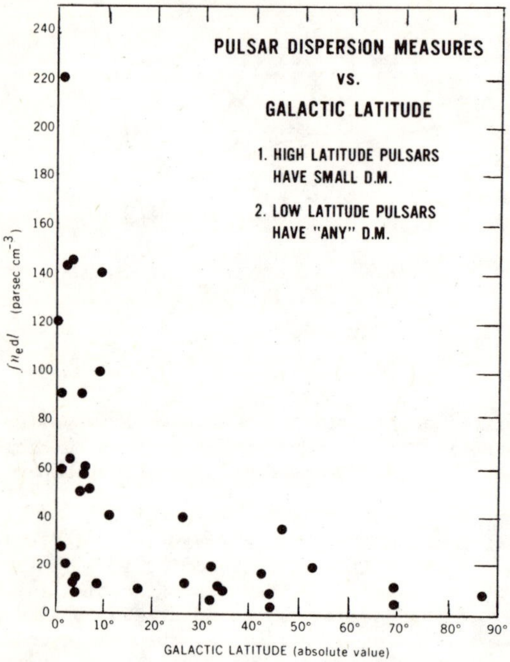

Fig. 3 - Pulsar dispersion measures versus galactic latitude.

The DM's have been analyzed by Mills (1969) in terms of a model for the interstellar electron gas in which N_e depends on distance normal to the galactic plane z according to

$$N_e(z) = \frac{N_e(o)}{1 + (z/a)^2} \tag{3}$$

Using this approach to the interpretation of the disperion measures of three distant pulsars, the quantity $N_e(o) \cdot a$ was found to be about

12 pc cm^{-3}, in agreement with expectations.

Observations of the interstellar absorption of 21 cm radiation gives the hydrogen measure:

$$HM = \int N_H \, ds,$$

proportional to the density of neutal hydrogen atoms, N_H. The ratio DM/HM leads to the degree of ionization and thus to a determination of the temperature of the interstellar medium, another very useful application of pulsar observations.

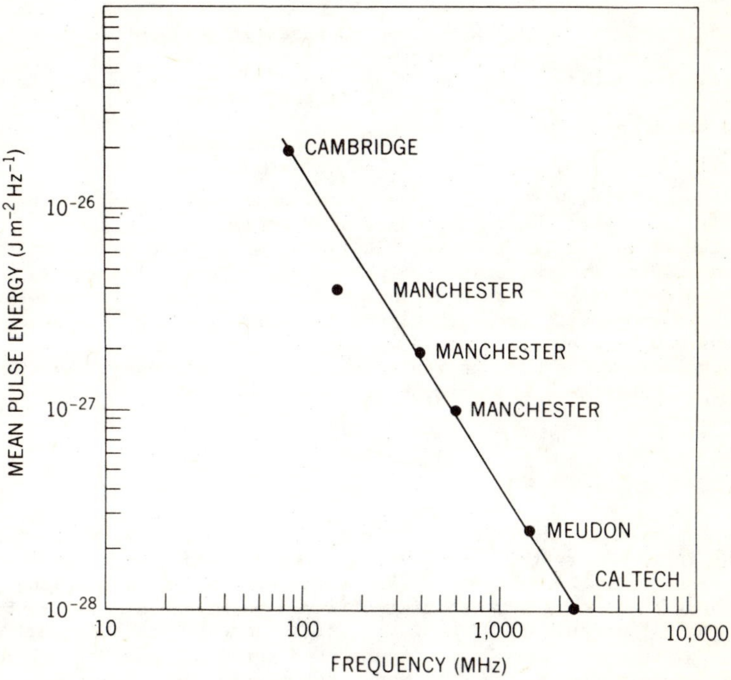

Fig. 4 - Radio spectrum of CP 0808, obtained by combining observations at several observations and thus employing disparate techniques.

Figure 4 shows the radio spectrum of CP 0808. The nice linearity of this spectrum is, however, not typical. Most pulsars have high frequency cut-offs, and some have low frequency cut-offs as well. The spectrum of NP 0532 extends from the radio to the X-ray regions, and shows two distinct components (see Figure 1 of Pacini's paper in this volume). There is no lack of theories that can account for two components; the problem is to account for the high brightness temperature of the radio component.

Pulsar repetition rates vary from about 30 per second to about one per 3.7 seconds. In all cases of measured period changes, the pulse rate is slowing down. The single exception to this general rule is the extraordinary event that occurred in the pulsar in the Vela X supernova remnant (PSR 0833-45) in early 1969. The period abruptly decreased by about one part in 10^6 and then resumed increasing at a rate that was about one percent greater than it was before the speed-up (Reichley and Downs 1969). Several ingenious theories were proposed to explain the speeding up. M. Ruderman suggested that the outer layer of a neutron star is crystalline and capable of supporting a certain amount of shear. Solidifying at an early epoch, while the neutron star is rotating much faster than the pulsar's present rate, the crust will be oblate due to centrifugal force. As the star slows down the equilibrium shape changes and the stresses on the crust increase until it cracks and collapses. In this "starquake" the moment of inertia is reduced and the star spins faster. Other theories include that of F. Dyson who proposes that hot, high-pressure material from the interior escapes through a hole in the crust, forming a "volcano." The released material then piles up around the volcano until its weight causes the crust to collapse. G. Greenstein and A.G.W. Cameron have proposed still another theory, which relates the Vela "jump" to a hydrodynamic instability.

The difficulty in finding new, more distant, pulsars is that interstellar dispersion, which increases with distance, spreads out a pulse in time due to the finite receiver bandwidth, making successive pulses overlap and wiping out the pulsation phenomenon. Dispersion removal techniques, in which delays are introduced into different parts of the receiver bandwidth, to compensate for high dispersion, give promise of leading to the discovery of new pulsars. In this series we will be hearing about such new developments as well as reviewing all the past observations and all the major theories of pulsars and related phenomena.

Postscript

The preceding article is the edited, slightly abbreviated transcript of the introductory talk in the Physics of Pulsars seminar series. It was presented on 22 September 1969 at the University of Maryland. As this book goes to press, the list of known pulsars has risen to 55, but the Crab Nebula pulsar, NP 0532, is still the only known optical pulsar. A new instrument used by Chiu, Lynds, and Maran to search for optical pulsars was tested on the Crab and obtained the series of photos shown in Figure 5. The theoretical picture is still clouded, but the newer theories involving pulsar emission at a distance from the stellar surface describe the emission region as lying mostly significantly closer to the star than the velocity of light cylinder. Perhaps the area of greatest interest is the study of fluctuations in the pulse arrival times, including both brief discontinuities in the pulse repetition frequency and the fluctuations, on a time scale of months, that were initially described as quasi-sinusoidal. The theorists in this area are applying the physics of superconductivity and solid state phenomena in attempts to understand the internal constitution of neutron stars.

Approximate parameters for the 55 pulsars announced as of September 1970 are given in the Appendix.

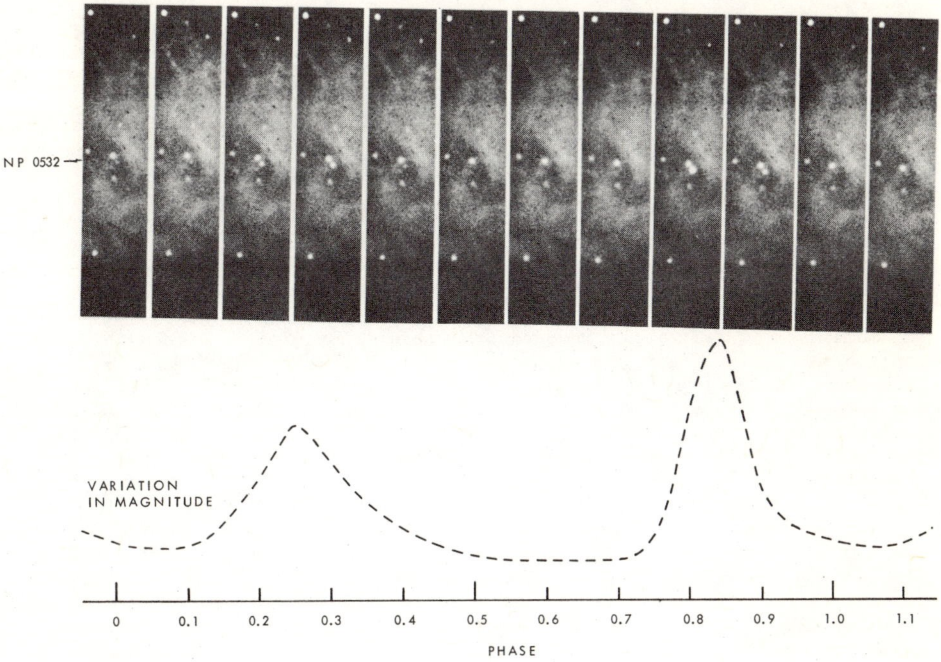

Fig. 5 - Phase resolved electronic image tube photos of the Crab Nebula pulsar, NP 0532, obtained with the 84-inch telescope on Kitt Peak.

REFERENCES

Chiu, H. Y. and Occhionero, F. 1969, Nature, 223, 1113.

Gold, T. 1968, Nature, 218, 731.

Gunn, J. E. and Ostriker, J. P. 1969, Nature, 221, 454.

Hewish, A., Bell, S. J., Pilkington, J. D. H., Scott, P. F., and Collins, R. A. 1968, Nature, 217, 709.

Michel, F.C. and Tucker, W. H. 1969, Nature, 223, 277.

Mills, B. Y. 1969, Proc. Astron. Soc. Australia, 1, 176.

Ostriker. J. P. and Gunn, J. E. 1969, Astrophys. J., 157, 1395.

Pacini, F. 1967, Nature, 216, 567.

________ 1968, Nature, 219, 145.

Radhakrishnan, V. and Cooke, D. J. 1969, Astrophys. Lett., 3, 225.

Reichley, P. E. and Downs, G. S. 1969, Nature, 222, 229.

Wheeler, J. A. 1966, Ann. Rev. Astron. Astrophys., 4, 393.

September 22, 1969

Revised October 1, 1970

RADIO OBSERVATIONS OF PULSARS

G. Richard Huguenin

University of Massachusetts
Amherst

Contrary to the announced title of this talk, it is not my intention to review every aspect of the radio observations of pulsars. I wish rather to concentrate this review to a discussion of amplitude variations, polarization, and spectra. It is my understanding that other aspects of radio observations, such as accurate timing measurements and searches for new pulsars, will be covered by subsequent speakers.

Certainly the most striking observational characteristic of the pulsar phenomenon is the pulsed nature of the radio emission. This feature undoubtedly has been reviewed by Dr. Maran. The precise timing of these pulses is to be discussed later on in this series. Perhaps the second most distinguishing characteristic of pulsar emissions at radio wavelengths is the large range of amplitude variations present. These amplitude variations occur with time scales ranging from tens of microseconds to months. Some of the amplitude variations are now known to be imposed by interstellar scintillations, and are not therefore associated with the source. The remainder of the variations, however, are believed to be intrinsic to the source region and are therefore related to the pulsar phenomenon.

Of primary importance in deducing an emission mechanism for the pulsars is a knowledge of the spectrum and polarization of the observed emission. When the original paper by Hewish, et al. (1968) appeared in Nature announcing the discovery of CP 1919+21, we decided that this source was indeed a fascinating object and immediately applied for observing time on the N.R.A.O. 300-foot transit telescope at Green Bank. Spectra and polarization seemed to be the obvious parameters for the first observations to be attempted with the transit telescope. We thus designed and built a polarimeter to operate over more than an octave between 65 and 165 MHz. From observations made during March and April 1968 it was found that the emission was highly polarized and generally elliptical, but that the polarization could vary significantly from pulse to pulse. At that time, data acquisition facilities were inadequate and we did not have sufficient time resolution to resolve polarization across the pulse. Subsequent to these crude initial observations the complete polarization properties for several pulsars have been accurately determined.

The state of polarization of the radio emission from pulsars is observed to fluctuate from pulse to pulse as is indicated in Figure 1. We see from this figure that the polarization is generally elliptical and that the percentage polarization is quite high. Although the polarization properties of the radio emission vary considerably from pulse to

Fig. 1. Polarization parameters for individual strong pulses for four different sources. The profiles shown are: _____ total intensity; ----- circularly polarized intensity (positive if right-hand); ..... linearly polarized intensity. Values are only shown where the intensities are greater than the peak noise. The error in the angle of the polarization ellipse varies across the pulse; a full line is used where the error is < 10°, a broken line if the error is between 10° and 30°, and no value is given for larger errors. Two pulses, (a) and (b), are shown for each source to indicaté pulse to pulse changes. CP 0328: (a) and (b) Nov. 1, 1968, with a 1 ms time constant. AP 0823: (a) Nov. 1, 1968, with a 0.1 ms time constant, (b) Nov. 22, 1968, with a 1 ms time constant. PSR 0833-45: (a) and (b) Nov. 14, 1968, with a 0.1 ms time constant. CP 1133: (a) March 25, 1969, with a 1 ms time constant, (b) Nov. 9, 1968, with a 1 ms time constant. (from Ekers, R.D. & A.T. Moffet 1969, preprint 8, "Polarization of Pulsating Radio Sources")

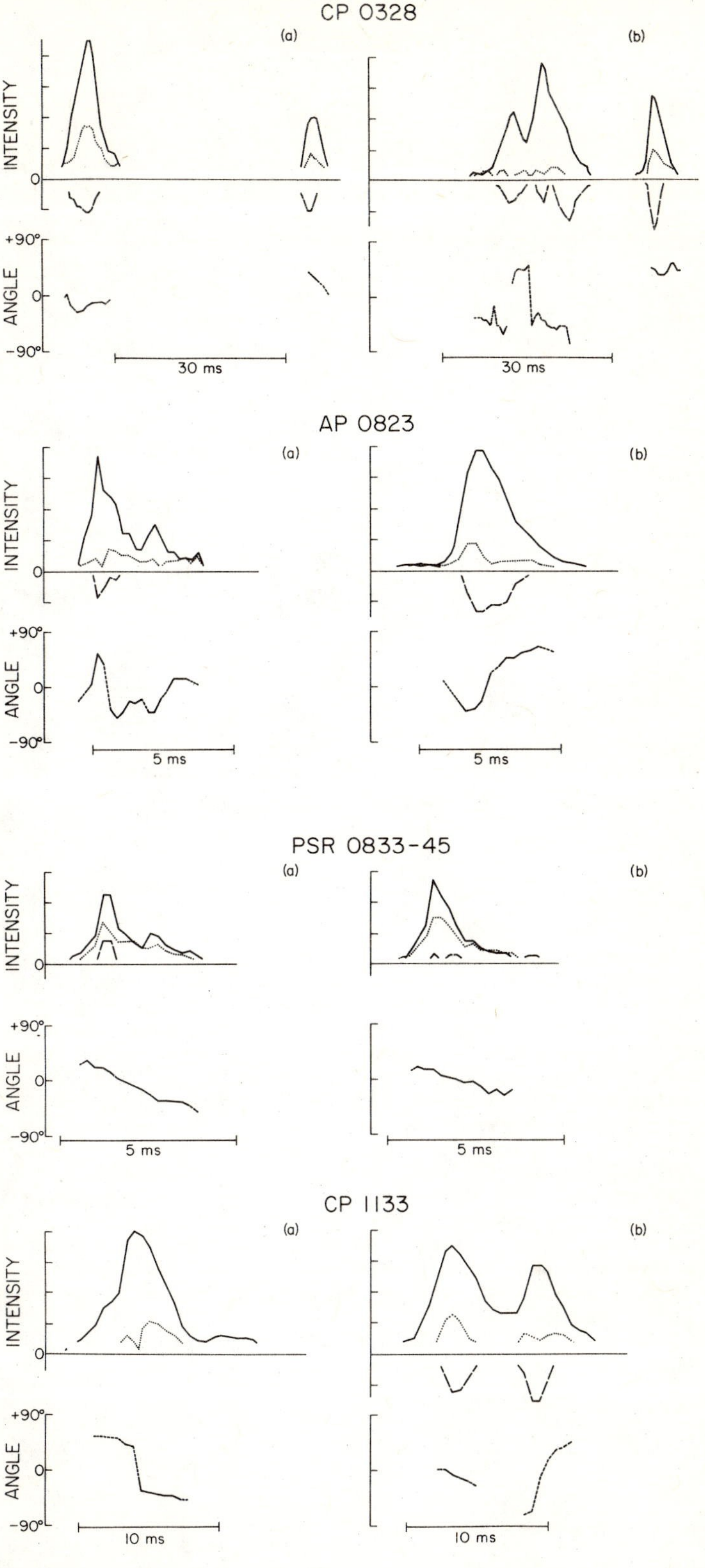

CP 0328
(a)
(b)
INTENSITY
0
ANGLE
+90°
0
-90°
30 ms
30 ms
AP 0823
(a)
(b)
INTENSITY
0
ANGLE
+90°
0
-90°
5 ms
5 ms
PSR 0833-45
(a)
(b)
INTENSITY
0
ANGLE
+90°
0
-90°
5 ms
5 ms
CP 1133
(a)
(b)
INTENSITY
0
ANGLE
+90°
0
-90°
10 ms
10 ms

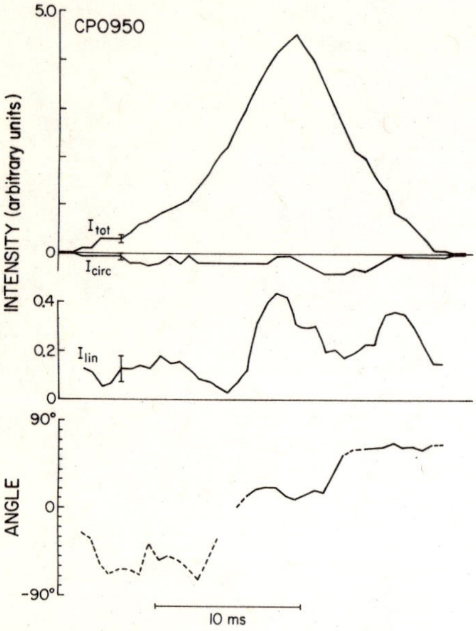

CP0950
5.0
INTENSITY (arbitrary units)
I_{tot}
I_{circ}
0
0.4
0.2
I_{lin}
0
90°
ANGLE
0
-90°
10 ms

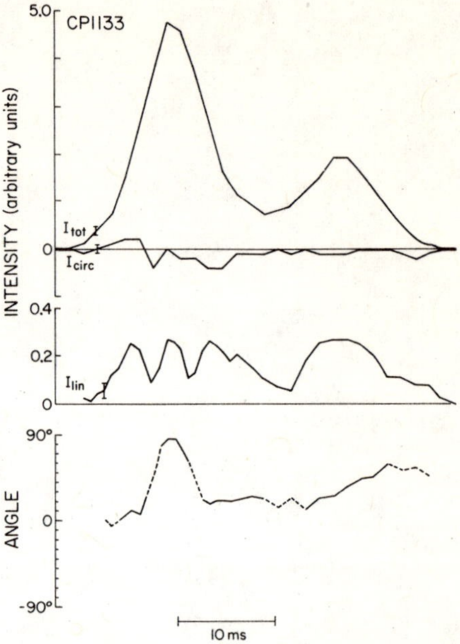

CP1133
5.0
INTENSITY (arbitrary units)
I_{tot}
I_{circ}
0
0.4
0.2
I_{lin}
0
90°
ANGLE
0
-90°
10 ms

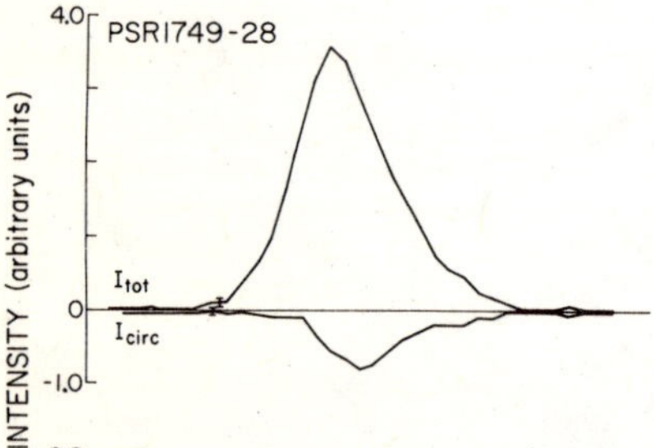

PSR1749-28
4.0
INTENSITY (arbitrary units)
I_{tot}
I_{circ}
0
-1.0
0.2
I_{lin}
0
10 ms

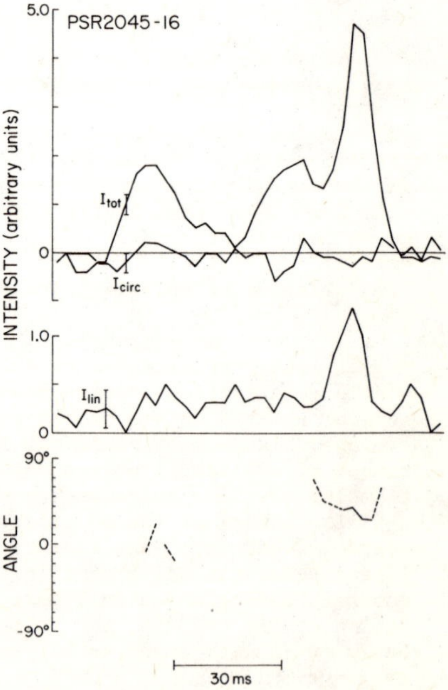

PSR2045-16
5.0
INTENSITY (arbitrary units)
I_{tot}
I_{circ}
0
1.0
I_{lin}
0
90°
ANGLE
0
-90°
30 ms

Fig. 2. Mean polarization parameters (15 minute average) for four pulsars observed on March 25, 1969, with a 1 ms time constant. The angle of the polarization ellipse has an arbitrary zero. Errors in the intensities are indicated by the bars near the beginning of each profile. The error in the angle varies across the pulse; a full line is used if the error is < 10°, a broken line if the error is between 10° and 30°, and no value is given for larger errors. (N.B., the linear intensity is plotted on an expanded scale.) (from Ekers, R.D. & A.T. Moffet 1969, preprint 8, "Polarization of Pulsating Radio Sources")

pulse, long-term averages of the polarization parameters indicate that a stable polarization average does exist. Thus, if we average together several hundred pulses of the type shown in Figure 1 we find that a stable average polarization state can be defined, i.e., an average of a like number of additional pulses results in the same net polarization curve. Examples of these averages are shown in Figure 2.

The nature of the polarization average varies considerably from source to source. Some features, such as 90° jumps in the orientation of linear polarization, appear in the average curves for several sources. Another feature shown in Figure 2 is the slowly rotating linear polarization of PSR 0833-45, the Vela pulsar. Although not unique, this behavior is not common to all sources.

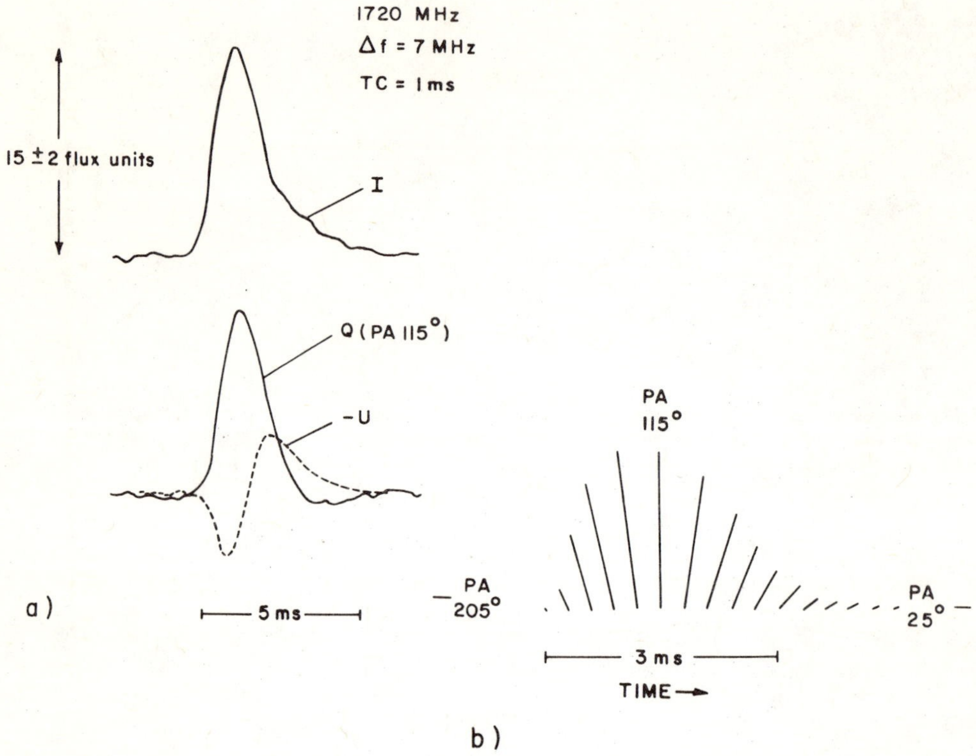

Fig. 3. The distribution of linear polarization across the average pulse from PSR 0833-45 (Vela) at 1720 MHz. a) The distribution of Q and -U represents a gradual change of approximately 45° in the position angle of polarization across the pulse. Within the errors of measurement, the pulse is completely polarized. b) Sketch showing the rocking of the plane of polarization during the pulse from PSR 0833-45. (from Radhakrishnan et al., 1969)

Figure 3 shows in a more vivid fashion much discussed rocking of the polarization across the average pulse from PSR 0833-45. Observed at 1720 MHz, this pulsar shows nearly 100% linear polarization with a continuously rotating position angle. Such polarization behavior is interpreted as strongly suggesting a rotating source. Models of rotating sources can likewise be constructed to adequately explain the observed 90° jumps. A discussion of such models will presumably be presented later on in the series. Additional observations are required to ascertain a more detailed behavior of the pulse-to-pulse variations in polarization.

The mean spectra for several pulsars have now been determined over a frequency range from about 40 MHz to 5 GHz. The Crab Nebula pulsar has also been observed in the optical and X-ray portions of the spectrum. Because of the wide range of time variations, mean spectra must be determined with great care. Figure 4 shows an early attempt to measure the pulsar spectrum of CP 1919+21 with each curve representing the spectrum of a single pulse. The pulse-to-pulse variations evident in Figure 4 are typical of pulsar behavior in general.

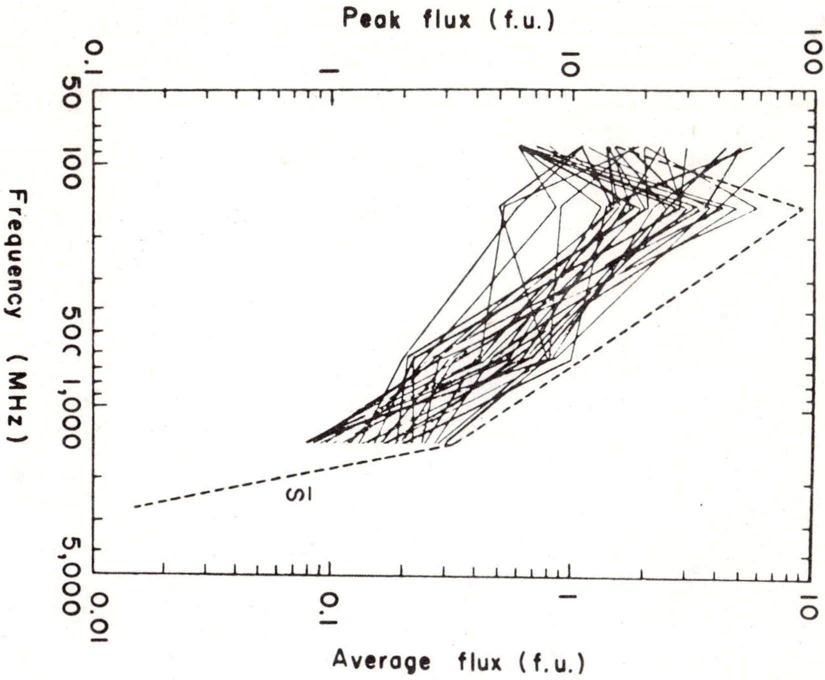

Fig. 4. Spectra of individual pulses from CP 1919+21 between 85 and 1410 MHz (peak flux scale refers to the continuous lines). The broken line shows the average pulse height on March 25, 1968 integrated over 20 minutes near 20^h sidereal time (average flux scale). The 2700 MHz measurement was made on March 28, 1968. (from Robinson et al., 1968)

In Figure 5 are shown the mean spectra of several typical pulsars. The main features common to all of these spectra are the steep cutoffs observed both at long and short wavelengths. The spectra of most pulsars fall off sharply at long meter-wavelengths. In fact, no pulsars have been observed at frequencies below 40 MHz (7.5 m) despite a considerable amount of searching at Clark Lake at 26 MHz (11 m) and at Penticton at 22 MHz (13 m). Pulsars are also intrinsically quite weak and difficult to detect at centimeter wavelengths.

Average pulse envelopes change shape with frequency. The spectral index appears to change across the pulse so that the relative amplitude and widths of subpulses change in a systematic way with frequency. This is demonstrated in Figure 6. Several sources, such as CP 1133+16, have subpulses whose widths seem to follow a dependence of $f^{-1/4}$. For the Crab Nebula pulsar, the pulse shape changes dramatically below 100 MHz, so that at 74 MHz, the pulse shape has deteriorated into a square wave.

The equivalent brightness temperatures derived for the radio emission from pulsars are extremely high. If we assume dimensions of the emitting region which are characteristic of the shortest scale time variations,

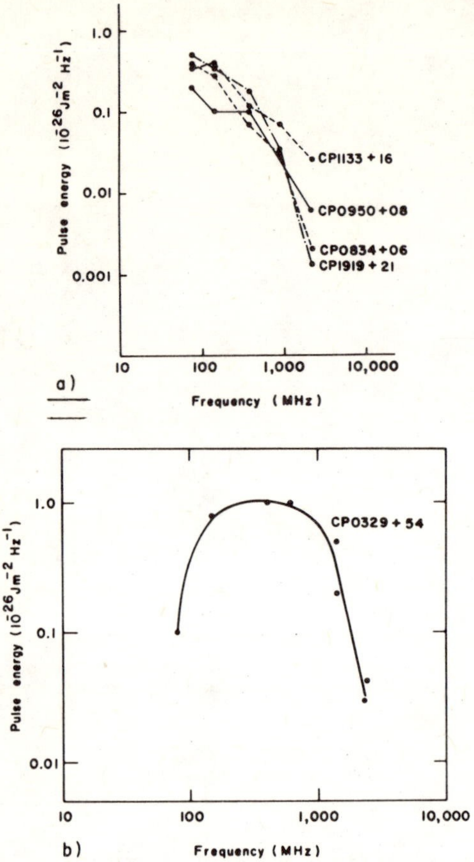

Fig. 5. Mean spectra of pulse energy from pulsars. a) for CP 0834+06, CP 0950+08, CP 1133+16, and CP 1919+21 (from Ekers, R.D. & A.T. Moffet, Nature 220, 756, 1968). b) for CP 0329+54 (from Maran and Cameron, 1969).

then peak brightness temperatures in the range of 10^{21} to 10^{26} °K are deduced. With brightness temperatures of those magnitudes, the emission mechanism must surely be a coherent process. For comparison, the highest brightness temperatures observed for solar radio bursts or for the quasi-stellar sources are 10^{10} to 10^{12} °K.

Dynamic spectra of pulsars have been obtained over a wide frequency range (100 to 400 MHz) in order to study the effects of interstellar scintillations on pulsar emission. We at the University of Massachusetts made usch observations with the 300-foot telescope at Green Bank employing a multi-channel spectrometer. The pulsar emission intensities in 50 adjacent frequency channels were recorded on 35 mm film every 5 ms. Examples of the spectra obtained with this instrument are shown in Figure 7. The band structure evident in Figure 7 is undoubtedly the result of interstellar scintillations. The de-correlation bandwidth (width of an average band in Figure 7) vs. center frequency for several sources is shown in Figure 8. The de-correlation bandwidth for a given source at a specified frequency is well correlated with the source's dispersion measure -- added proof of their origin in interstellar scintillations. The observed de-correlation bandwidths fit a power law relation of the form $\delta f \propto f^{-\alpha}$. The index α is observed to range between 2.3 and 3.6 for

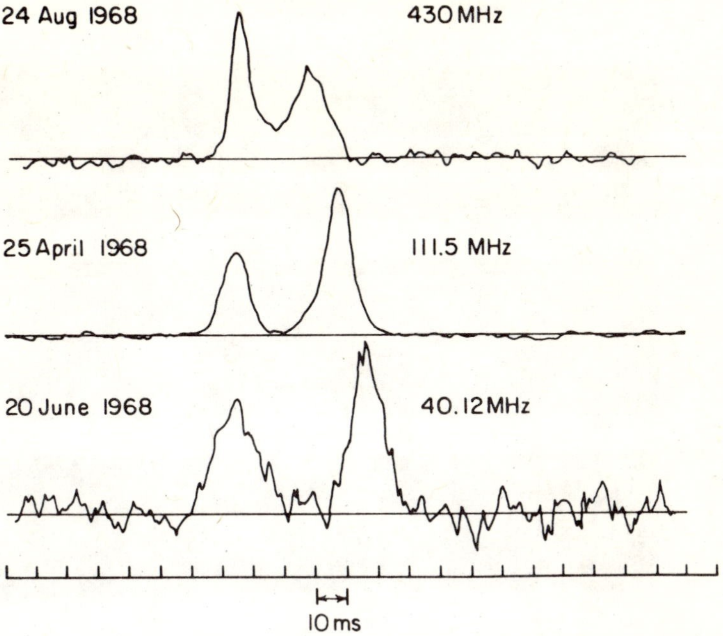

Fig. 6a. Average pulse shapes from CP 1133+16 as measured at three frequencies. The time scale at the base applies to all three pulses shown.

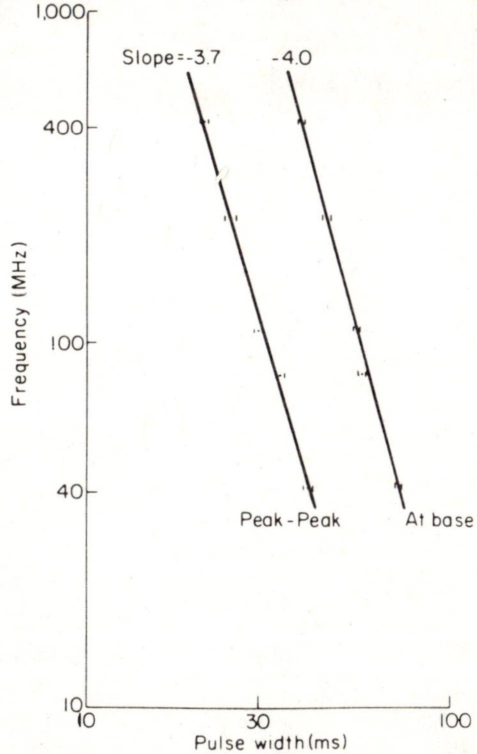

Fig. 6b. Average pulse widths versus radio frequency for CP 1133+16. The width noted as "peak-peak" is that measured between subpulse maxima. (from Craft and Comella, 1968).

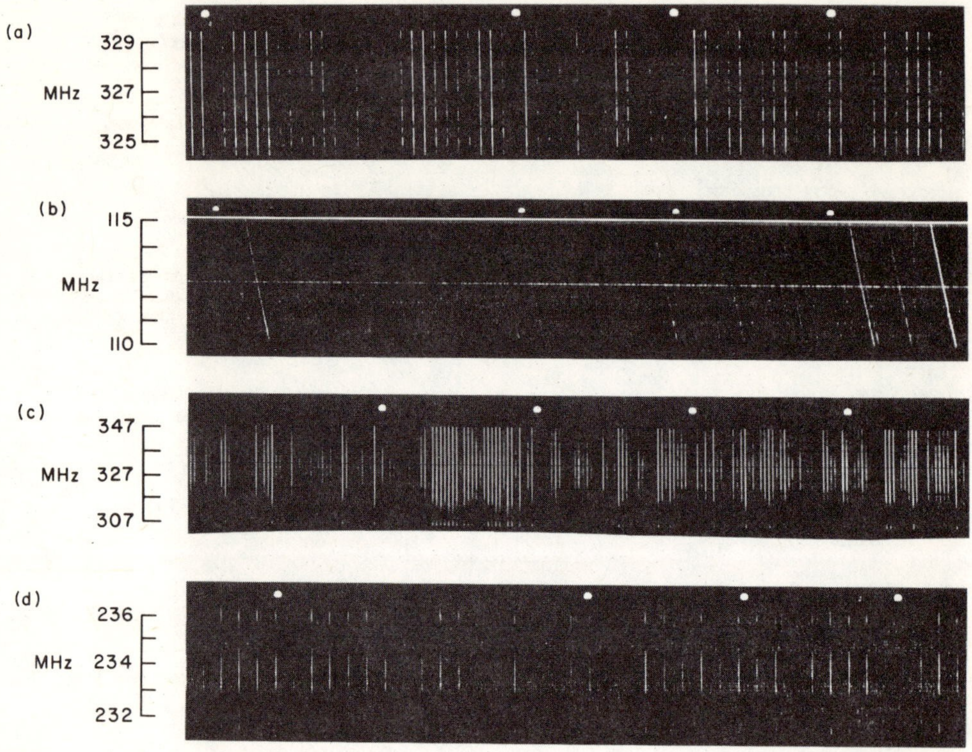

Fig. 7. Examples of dynamic spectra of pulsars; a) CP 0329+54, 27 Feb. 1969; b) NP 0531+21 (the Crab Nebula), 12 Feb. 1969; c) CP 0950+08, 28 Feb. 1969; d) CP 1133+16, 17 Feb. 1969. Frequency scales are given in MHz at the left of each record. Timing marks at the top of each film strip are separated by 10 sidereal seconds, with time increasing from left to right. (from Huguenin, Taylor, and Jura, 1969)

various sources whereas existing theories for scintillations predict $\alpha > 4$. Observations to further delineate the phenomenon are under way at various observatories. One can see from Figure 7 that long-term amplitude fluctuations will result from interstellar scintillations when observing bandwidths are on the order of, or narrower than, the de-correlation bandwidth. Interstellar scintillations are thus responsible for some of the observed amplitude variation, particularly in the range from several minutes to hours.

The amplitude variations with time scales ranging from the order of a second (one pulse period) to several minutes are broadband in nature and are presumably associated with the source. A careful analysis of these amplitude variations has, in fact, shown that these fluctuations are not random, but contain multiple periodicities. Such periodic phenomena must be associated with the source region.

The original data used for the analysis of pulse-to-pulse amplitude variations were obtained with the same dynamic spectrograph used for interstellar scintillation studies. The amplitudes of successive pulses

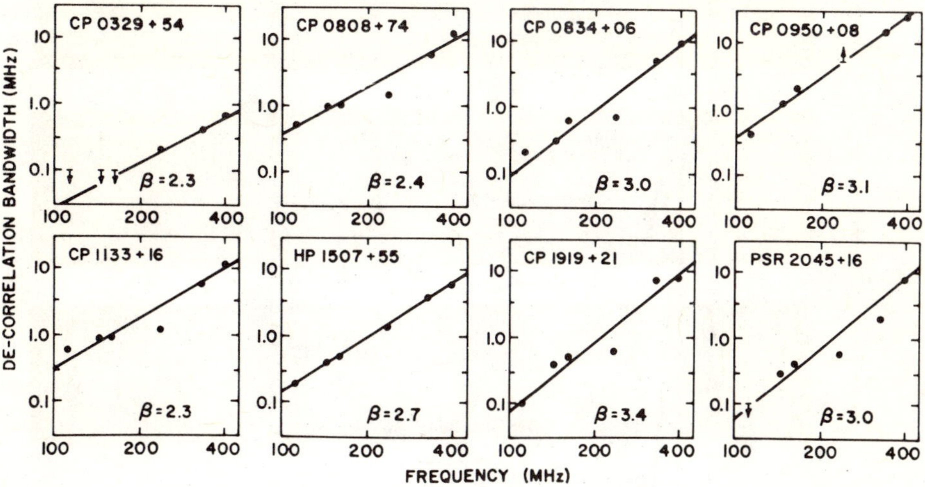

Fig. 8. De-correlation bandwidth plotted as a function of frequency for 8 pulsars. Straight lines represent power law fits to the data with index β. (from Huguenin, Taylor, and Jura, 1969).

TABLE 1

MODULATION FREQUENCIES OF 10 PULSARS

Pulsar	f_m (cycles per pulse period)	Q
MP 0031-07	0.012	1.7
CP 0329+54	0.1	1.0
CP 0808+74	0.090	9.0
CP 0823+26	0.18	2.0
CP 0834+06	0.463	29
CP 1133+16	0.035	2.5
AP 1237+25	0.35	8.8
PSR 1642-03	0.074	4.1
CP 1919+21	0.228	15
PSR 1929+10	0.094	3.1

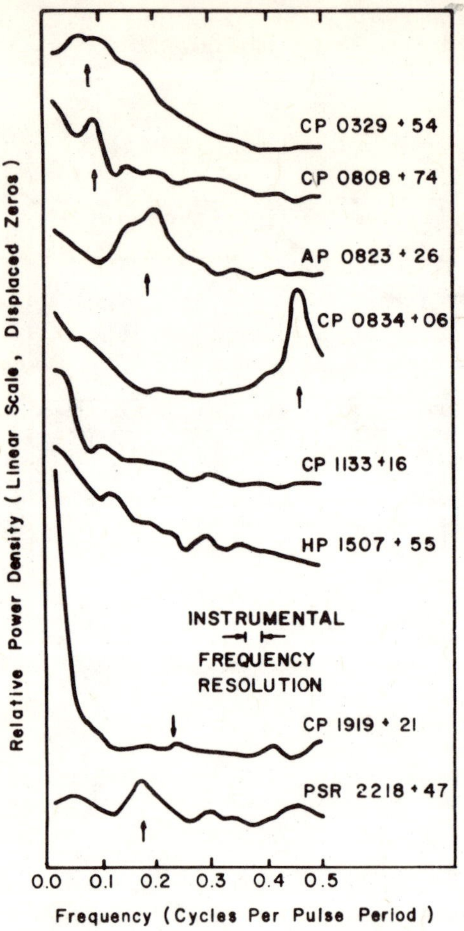

Fig. 9. Mean spectra of the pulse-to-pulse intensity fluctuations of eight pulsars. Frequencies are given in cycles per pulse period, and may be converted to Hz by dividing by the period in seconds. Significant features in the spectra are marked by upward-pointing arrows. The downward-pointing arrow marks the frequency of a feature observed in the fluctuation spectrum of CP 1919+21 by Lovelace and Craft. (from Taylor, Jura, and Huguenin, 1969).

were digitized and a power spectrum computed. The fluctuation spectra for several sources are shown in Figure 9. Multiple periodicities have been identified. These periodicities are similar to those found at Arecibo by Drake and Craft (Drake, F.D. & H.D. Craft, jun. Nature 220, 231, 1968) however we interpret these not as aliases of high frequency ($\sim$100 Hz) pulsations, but rather as periodicities with frequencies lower than the main pulsation period. High time-resolution observations are currently planned to try to identify the phenomenon responsible for the observed periodicities. In order to obtain the necessary signal-to-noise ratio for high time-resolution observations, a pulse compression technique is required. Such a receiver is currently nearing completion at the University of Massachusetts. We remain at a loss to explain such periodicities, however, at least until further observations are obtained. A list of observed periodicities is given in Table 1.

REFERENCES

Craft, H.D., Jr. and Comella, J.M. 1968, Nature, 220, 676.
Drake, F.D. and Craft, H.D., Jr. 1968, Nature, 220, 231.
Ekers, R.D. and Moffet, A.T. 1969, Ap. J. (Letters), 158, L1.
Hewish, A., Bell, S.J., Pilkington, J.D.H., Scott, P.F., and Collings, R.A. 1968, Nature, 217, 709.
Huguenin, G.R., Taylor, J.H., and Jura, M. 1969, Astrophys. Lett., 4, 71.
Maran, S.P. and Cameron, A.G.W. 1969, Earth and Extraterrestrial Sci., 1, 3.
Radhakrishnan, V., Cooke, D.J., Komesaroff, M.M., and Morris, D. 1969, Nature, 221, 443.
Robinson, B.J., Cooper, B.F.C., Gardiner, F.F., Wielebinski, R., and Landecker, T.L. 1968, Nature, 218, 1143.
Taylor, J.H. Jura, M., and Huguenin, G.R. 1969, Nature, 223, 797.

October 6, 1969

OPTICAL OBSERVATIONS OF PULSARS

E. J. Wampler

Lick Observatory
Board of Studies in Astronomy and Astrophysics

Since the first radio detection of pulsars some two and one-half years ago, there have been repeated attempts to find optical counterparts to these radio objects. Searches made near the radio positions of CP 1919, CP 0950, CP 1133 and PSR 0833-45 have been unsuccessful in detecting optical counterparts, although in the summer of 1968 there was some excitement when it was believed that the optical counterpart of CP 1919 had been discovered with a period equal to twice the radio period. Additional work failed to confirm these early results and it is now believed that the timing signals used to synchronize the electronic equipment introduced noise which produced spurious signals.

In order to convert the observed brightness of an object into an intrinsic brightness it is necessary to know its distance. This is always a difficult problem even in optical astronomy, and if the object is not seen optically, the difficulties are compounded. Distances are determined from the radio data by using the dispersion measure and an estimate for the interstellar electron density. Although the dispersion measure can be accurately determined, the estimates for the line-of-sight electron density are very crude and introduce large errors in the distances using the best distance estimates available. Kristian (1968) has shown that for CP 0950 and CP 1133 the absolute visual magnitude is fainter than 19 and 18, respectively, and for PSR 0833-45 Freeman, Rodgers, Rudge, and Lyngå (1969) have shown that the absolute visual magnitude is fainter than 12.

The apparent faintness of the optical counterpart of pulsars had lead some theorists to speculate that the optical counterpart of radio pulsars would never be found. This makes the discovery on January 16, 1969 of the optical counterpart of NP 0532 by Cocke, Disney, and Taylor even more striking. The apparent visual magnitude of the Crab Nebula pulsar (NP 0532) is about 16th mag and leads to an absolute visual magnitude of about 3rd magnitude. Therefore the difference in brightness between the Crab Nebula pulsar and the next brightest one amounts to a factor of about 10^4. Any theory that explains the optical emission, and at present there is no such theory, would have to explain this enormous difference in brightness.

How does one go about detecting pulsars optically? One method is to mount a photomultiplier behind a focal plane diaphragm. The diaphragm is used to exclude background light from all but a small region of the sky surrounding the suspected position of the pulsar. If the aperture of the diaphragm is increased, the ratio of the pulsar signal to the background noise decreases, but there is an increased tolerance to error in the position of the pulsar. The photomultiplier behind the diaphragm detects the starlight and the surrounding sky background and sends this combined signal to a "black box." The job of the "black box" is to analyze the detected signal and sort the fluctuating component of the light that is due to the pulsar from the sky background. This method has the advantage that with Fourier Transform techniques one can look at all frequencies simultaneously. It has the disadvantage that the position of the pulsar must be rather well known.

A second method involves the use of a chopper and some form of area detector such as a photographic plate or TV camera. The function of the chopper, which may be either a mechanical device or an electronic gate, is to gate the detector with the same period as the pulsar. If the chopper is in phase with the light flashes the pulsar will be seen, otherwise it will disappear while field stars will remain with a constant brightness. By chopping in synchronization with the pulsar, much of the background radiation can be eliminated and comparatively faint signals can be detected. Also one can cover a large area of the sky simultaneously. This technique has a disadvantage in that it is tedious to conduct searches in frequency space because one is tied to the chopper.

The Crab Nebula pulsar was discovered by Cocke, Disney, and Taylor (1969) using a fairly large diaphragm to locate the pulsar to within about 5 arc-seconds of the central region of the Crab Nebula. The center of activity of the Crab was known to lie in this area and as early as 1942 Minkowski, on the basis of spectroscopy and proper motions, had

Fig. 1. Two pictures showing the Crab Nebula pulsar with three field stars. The upper photograph shows the pulsar near maximum light and the lower photograph shows it near minimum.

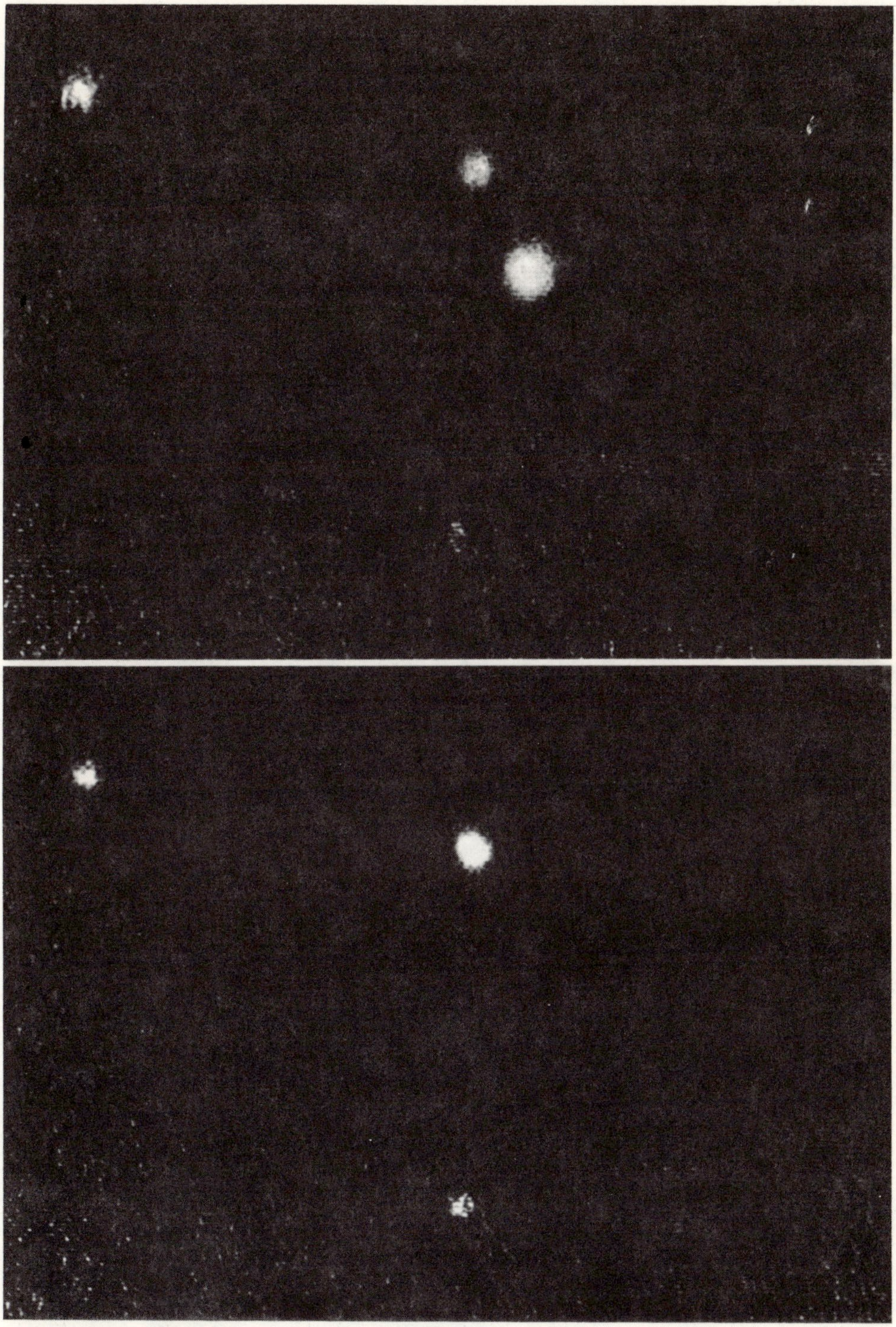

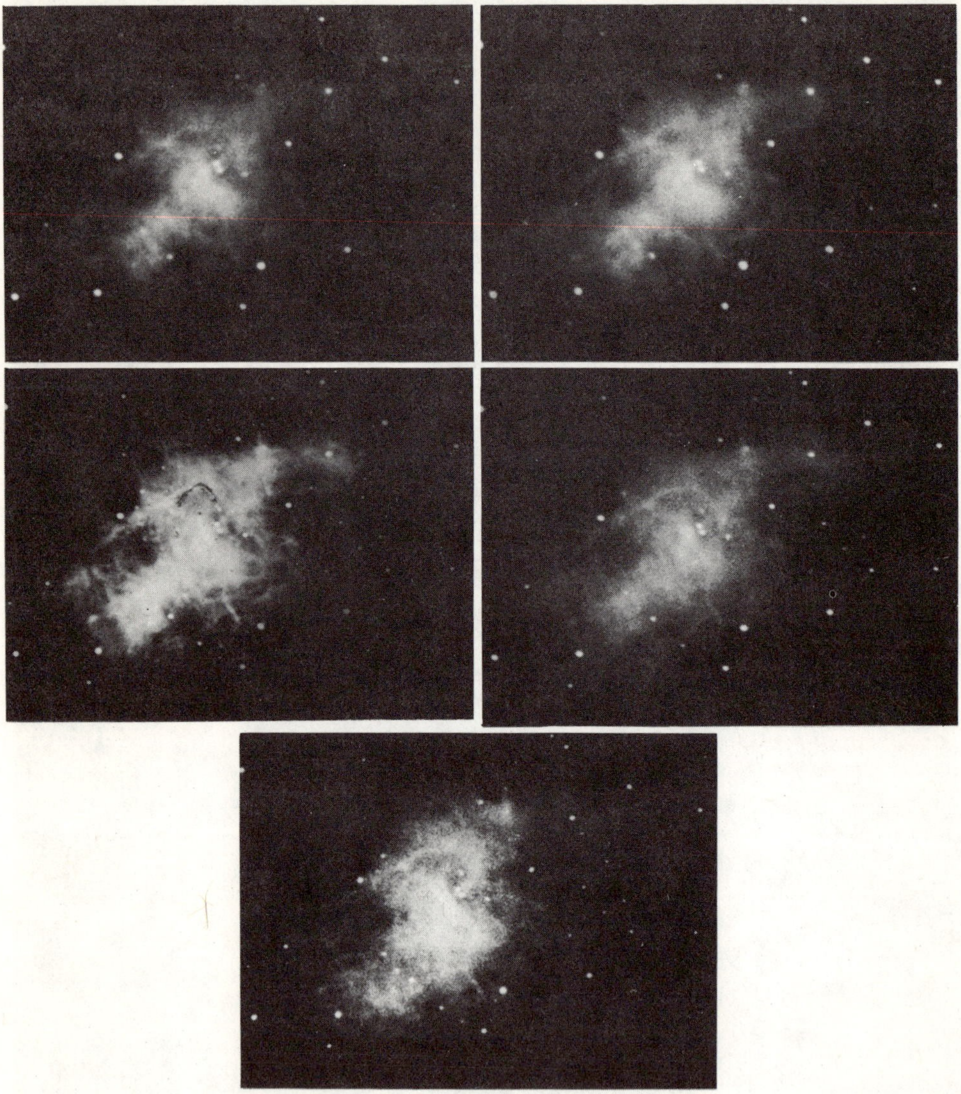

suggested that one of the stars in this region was the supernova remnant. Later Scargle (1969), by examining the motions of several light "wisps," concluded that there was some evidence that the center of activity was approximately 1 arc-second west of Minkowski's star. Following the discovery by Cocke, Disney, and Taylor a number of investigating groups turned their attention to the Crab Nebula pulsar. This talk is concerned primarily with the results of these investigations from the time of discovery until the Crab Nebula became unavailable for nighttime work in the late spring of 1969.

Minkowski's star was soon identified as the pulsar. Lynds, Maran, and Trumbo (1969) reduced the positional uncertainty of the pulsar to $\pm$ 0.5 arc-seconds and concluded that Minkowski's star was the pulsar. Miller and Wampler (1969) using a TV camera and a mechanical light chopper, independently showed that Minkowski's star flashed on and off with the period of the pulsar (Figure 1). Because it is now known that the image of Minkowski's star seen on ordinary photographs is produced by the pulsar, it is possible by comparing plates taken near the turn of the century with modern plates, to investigate possible secular changes in the light output. Figure 2 shows plates taken in 1889, 1909, and 1914 together with a plate taken in 1969. There is some indication that the pulsar was slightly brighter on the earlier plates, but any difference in brightness is less than a factor of about two. These plates cover a period of time corresponding to roughly 1/10 of the lifetime of the pulsar and they set some limits to the decrease in optical brightness of the object.

Fig. 2. Four early photographs of the Crab Nebula together with a photograph (lower) taken in Sept., 1969. The top row shows two Crossley plates taken in 1899. The second row shows a picture (left) taken by Richie in 1908 using the Mt. Wilson 60-inch telescope; the right-hand picture was taken at the Crossley in 1914.

The radio emission from the Crab pulsar is known to vary with a time scale of the order of several minutes, and observations searching for similar effects at optical wavelengths have been made. Duthie, Sturch, Richie, and Rodnay (1969) reported that the ratio of the main pulse to the secondary pulse varied from 2.3 to 3.8 in a time scale of a few minutes. Similarly, Freeman, Rodgers, Rudge, and Lyngå (1969) also reported variations of this nature. However, Warner, Nather, and MacFarlane (1969), using a larger telescope, have failed to find these short-period variations. Results obtained at Lick Observatory by Wampler, Scargle, and Miller (1969) agree well with the data given by Warner et al. and fail to show variations similar to those reported earlier. The Lick results, which are presented as average curves after summing between 4,000 and 10,000 periods, are shown in Figure 3.

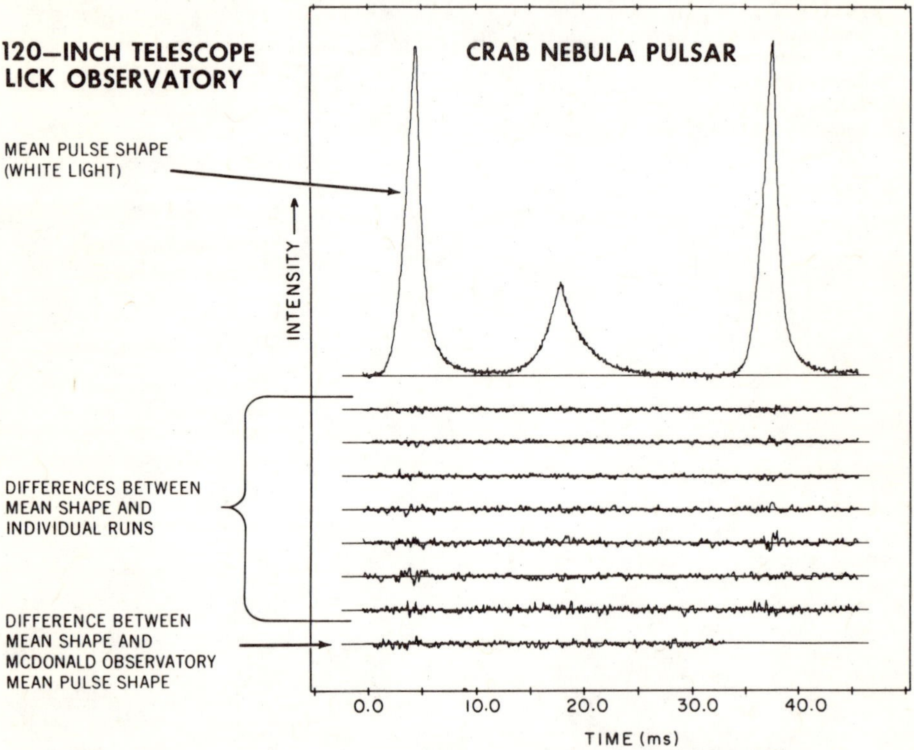

Fig. 3. Upper curve is the mean light curve in white light. The peak intensity occurs at 4.43 msec. Lower curves are differences between individual runs and the mean curve. Intensities for the individual runs were normalized so that the integrated intensity of the first main pulse was equal to that under the first main pulse of the mean curve. Data for the first three difference curves were obtained on March 14, 1969, at 3^h48^m, 4^h07^m, and 4^h17^m U.T., respectively. The next two curves were obtained on March 16 at 3^h47^m and 3^h52^m U.T., and the next two curves were obtained on March 23 at 3^h38^m and 3^h43^m U.T. The last curve is the difference between our mean curve and that published by Warner, Nather, and MacFarlane (1969).

The curves were plotted by a computer by requiring the pen to move from one data point to the other while remaining in contact with the paper. The fluctuations that are seen were in the curve are caused by statistical noise in the signal.

Workers at both Lick Observatory and Mt. Palomar have investigated fluctuations on a shorter time scale in the strength of single pulses. Kristian (1969) in examining a sample of the data obtained with the 200-inch telescope concluded that the pulse-height distribution observed in 200 successive main pulses was close to the distribution to be expected from photon-counting statistics alone. Drs. David Cudaback and Jared Anderson obtained extensive records on magnetic tape at Lick Observatory and are now in the process of analyzing their data. The data do not show any gross changes in pulse-to-pulse amplitudes, but a definitive result will have to wait for the completion of the numerical analysis. In any case, the large variations seen in the pulse-to-pulse amplitudes of the radio pulses are not seen in the optical data.

Several groups, by monitoring the arrival time of pulses, have investigated the spatial relationship of the regions responsible for the radio, optical and x-ray emissions. At Lick Observatory, Conklin, Howard, Miller, and Wampler (1969) established a radio link between the 120-inch Lick telescope and the 150-foot Stanford radio telescope and

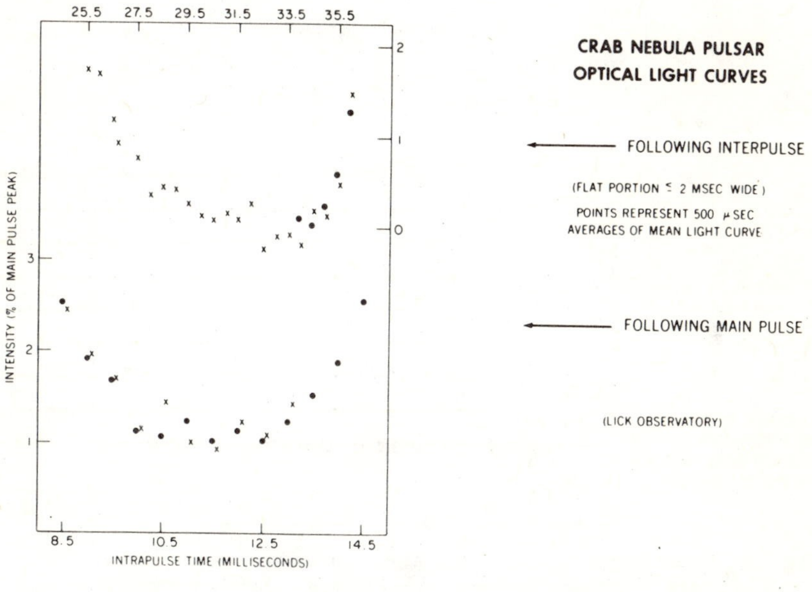

Fig. 4. Light curves for the intervals between the two peaks. The zero point of the time axis is the same as that in Fig. 3. Zero intensity was defined by the mean intensity in the 32 - 33.5 msec interval as shown in the upper curve. Each point is a 500-μsec average of the data. Dots and crosses are from the first and second halves of the light curve, respectively. Upper points are taken from the interval between the second and main pulse, lower points from the interval between the main and second pulse.

compared the arrival time of the optical pulses with the 424 MHz radio pulses. When the value for the interstellar dispersion was folded into the results, it was found that to the accuracy of the published dispersion measure the optical and radio pulses occurred simultaneously. A more accurate value for the dispersion obtained by Goldstein and Meisel (1969) has strengthened this conclusion. The new data shows that the difference in arrival time does not exceed 0.4 msec; this corresponds to a physical dimension of 120 km. Bradt et al. (1969) have accurately timed the arrival of x-ray pulses and compared them to the arrival of the optical pulses. They find that the pulses arrive simultaneously with an uncertainty of about 1 ms. This means that to within the observational uncertainties of a few hundred km, the radio-emitting region, the optical-emitting region, and the x-ray-emitting region are coincident.

Figure 4 shows the interpulse region. It is clear that between the main pulse and the interpulse the intensity does not drop to the level seen just before the main pulse. Other photoelectric data show that just before the main pulse the pulsar is very faint, but the data shown in Figure 4 indicate that there is no substantial interval in the light curve that is of constant intensity, so there may be some residual intensity at all times.

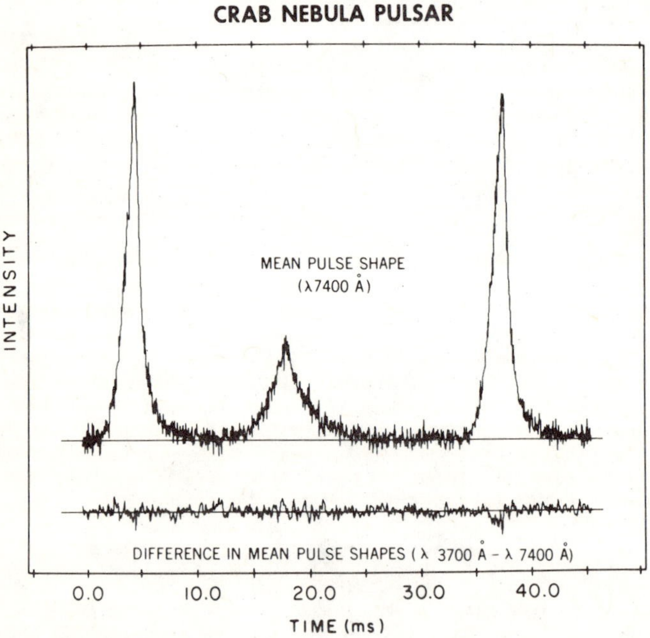

Fig. 5. Above: Mean light curve obtained in the infrared (7400 Å). Below: Difference between the mean light curve in the ultraviolet (3700 Å) and that in the infrared. Before the curves were subtracted, they were normalized by equating the areas under the first main peak. The lower curve has the same vertical scale as the upper curve.

Warner, Nather, and MacFarlane (1969) found essentially no color dependence in the shape of the pulses, using filters which approximated the U,B,V color system. Lick data are shown in Figure 5. These data were obtained using an S-20 photocell and cover a wider wavelength band, but essentially give the same null result found by Warner et al. There is a slight hint in the Lick data that the leading wing of the interpulse is slightly blue, but at the present time this should probably be regarded as an indication of the uncertainty of the data. It is interesting, however, that the x-ray data of Bradt et al., (1969) show a relatively intense interpulse leading edge.

The optical energy distribution of the Crab Nebula pulsar has been obtained by Oke (1969) and by Neugebauer et al., (1969). Their data indicate that in the frequency range 10^{14} - 10^{15} Hz the spectrum shows approximately the same distribution as a black body of 10^4 °K. However, if the volume of the emitting region is given by (c x pulse duration) then the energy emitted at 2.2μ exceeds that emitted by a 10,000° K black body by a factor of 10^6. Care must be exercised in fitting detailed models to the observed curves because the interstellar reddening correction is very uncertain. It is certain, however, that the overall energy distribution of the pulsar and the nebular background is completely different.

Polarization data has been obtained by several groups but only observers using large telescopes have obtained definitive results. Polarization results obtained at Lick Observatory using the 120-inch telescope are in qualitative agreement with the results obtained by Warner, Nather, and MacFarlane (1969) using the 82-inch telescope at McDonald Observatory, although there were substantive differences in the amount and direction of the polarization found by the two groups.

The Lick data (Wampler, Scargle, and Miller, 1969) were obtained using a single-channel polarimeter. The data were collected on two different nights and consisted of a dozen integrations with a Polaroid set at 30° intervals. During subsequent examination it was found that fluctuations of approximately 10% occurred for identical Polaroid settings. This fluctuation is larger than can be accounted for by the "shot" noise in the signal, and was larger than we would have expected from our subjective estimate of the quality of the night, although thin clouds may have been present. The formal average of the polarization, taken over a complete cycle, is 6.5% ± 0.9%. The error extimate was obtained from the observed scatter in the individual observations.

Fortunately the variation in the polarization can be obtained in a manner that does not depend on the extinction changes. Let the measured light curve be denoted by $I(\theta_i,t)$ where θ_i is the position angle and t is time, in terms of the Stokes parameters. Writing the Stokes parameters as a vector $(S_i) = (I,Q,U)$ we put

$$I(\theta_i,t) = H_{ij} S_j \quad (1)$$

where

$$H_{i1} = \frac{1}{2}, \qquad H_{i2} = \frac{1}{2} \cos 2\theta_i, \qquad H_{i3} = \frac{1}{2} \sin 2\theta_i \quad (2)$$

then

$$S_i(t) = H_{ij}^{-1} \; I(\theta_j,t). \tag{3}$$

Define the normalized light curves by

$$I^*(\theta_i,t) = \frac{I(\theta_i,t)}{\int I(\theta_i,t)dt} = \frac{I(\theta_i,t)}{<I(\theta_i)>} \tag{4}$$

where the integral is taken over the entire period of the pulsar. If S_i^* (t) is defined by

$$S_i^* (t) = H_{ij}^{-1} \; I^*(\theta_j,t) \tag{5}$$

then (3) can be written

$$S_i(t) = H_{ij}^{-1} \; H_{jk} <S_k> I^*(\theta_j,t) \tag{6}$$

where $<S_k>$ are the Stokes parameters giving the average polarization.

Now, since the average polarization is of the order of 6%, an important simplification is possible, namely if

$$<S_2> \; << \; <S_1>,$$

and

$$<S_3> \; << \; <S_1>,$$

then (6) reduces to

$$S_i(t) = \frac{1}{2} <S_i> \; S_i^*(t). \tag{7}$$

Thus the true Stokes parameters are proportional to the normalized ones and the true position angle $\theta = 1/2 \tan^{-1}(S_3/S_2)$, and the degree of polarization, $p = (S_2{}^2 + S_3{}^2)^{1/2}/S_1$, are identical with those caluclated from S_i^*, which is not very sensitive to the atmospheric transmission. The error made in this approximation is less than $|<S_2> + <S_3>/<S_1>|$.

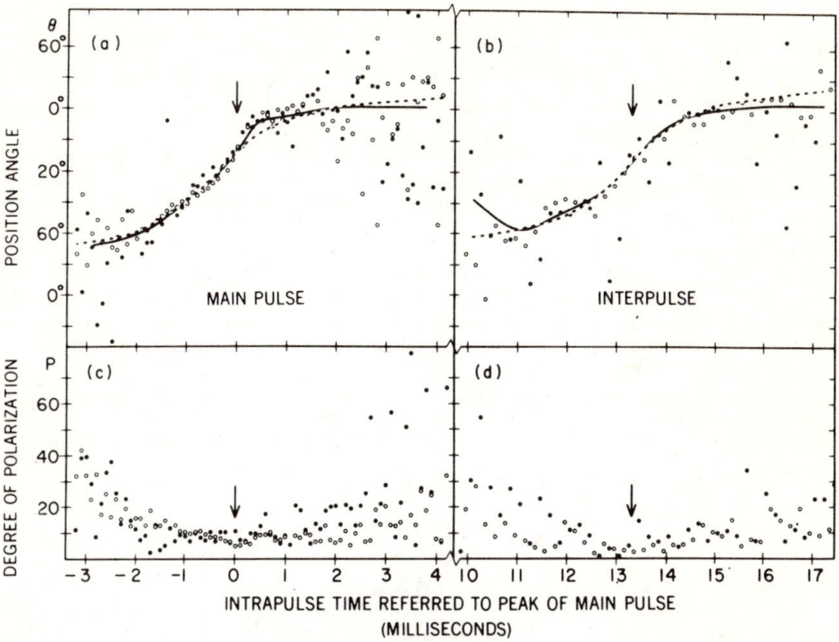

Fig. 6. Polarization as a function of time for the main pulse (a and c) and secondary pulse (b and d). Open symbols are for March 16, and closed symbols are for March 23 (signal-to-noise ratio was higher on the former night by about a factor of 4). Arrows mark points of maximum intensity (defined as t = 0 for the main pulse). Upper curves give position angle of vibration of the E-vector, and lower curves give degree of polarization.

Figure 6 shows the polarization data. The data from the main pulse are given on the left and the data on the right are for the secondary pulse. The polarization rises in the wings but one must be careful because, since degree of polarization is a positive definite quantity, scatter in the data will cause an apparent rise in the emergent polarization. The lack of points near zero polarization indicates, however, that this rise is real.

To explain the polarization data we consider a rotating star model. As early as March, 1968 Dr. J.J. Monaghan, in an unpublished work, had proposed a pulsar model with a rotating neutron star as the basic timing element and using an interaction between a strong magnetic field and atmospheric electrons to produce the observed radio emission. Subsequently other theoreticians have arrived at similar conclusions. In Figure 6 the solid lines are running (Stokes) averages of the data, and the dashed curves are theoretical curves based on the following model. Suppose that the source of radiation is fixed to an object which is rotating at the pulsar frequency about the rotation axis and there is another axis or radius vector in the source. Suppose also that the direction of polarization is perpendicular to the projection of this radius vector, called the polarization axis, on the plane of the sky. Now you can convince yourself that the amount of rotation of the direction of polarization during the cycle depends on the latitude of the polarization axis. The polarization position angle of the observed radiation is given by

$$\tan \theta = (\sin i \cot \beta - \cos i \cos \omega t)/\sin \omega t \qquad (8)$$

where i is the angle between the rotation axis and the direction to the observer, β is the angle between the rotation and the polarization axis and ωt is the phase of the rotation, taken to be zero when the polarization axis is closest to the observer. We have chosen the notation commonly used by workers in the field of magnetic variable stars.

Because the observed curve is only well-determined for small values of ωt the term in $\cos \omega t$ doesn't really contribute very much to fitting the curve and one essentially has a one parameter curve to fit, that parameter being

$$\sin i \cot \beta - \cos i.$$

Of course this parameter has two angles that must be adjusted.

Now it turns out that for the Crab pulsar the same curve will fit both the main pulse and the interpulse quite well. This allows one to say something about β. Suppose that the second pulse is formed when the other end of the polarization axis comes around. Then the latitude for both poles is the same, i.e., $\beta = 90°$ and the polarization axis is just about perpendicular to the rotation axis. Incidentally, it turns out that one can tell very easily that the line of sight does not pass between the two poles of the polarization axis because in this case the direction of rotation of the two pulses would be opposite.

We generated a variety of curves on the computer and fitted them to the data by eye, and we found that the best fit was for $\beta = 90°$ and $i = 77°$. However, in doing this we allowed a 10° vertical shift between the curves for the main pulse and the interpulse. There shouldn't be such a vertical shift because the position angle of the polarization at the inflection point of the curve is perpendicular to the rotational axis of the star, and must be identical for both pulses. Therefore we added the constraint that both pulses must give the same inclination of the axis. With this constraint and using a least-squares fit, $i = 76°$ and $\beta = 91°$. The uncertainty in these angles is a few degrees.

We also measured the circular polarization using a quarter wave plate. We found practically zero circular polarization. The formal results were 3.9% ± 4.9% for the integrated curve.

So we find that the polarization data can be fit by a model in which the polarization axis is very nearly perpendicular to the rotation axis. This polarization axis can be plausibly identified with the magnetic axis that is hypothesized in theoretical models of pulsars. The rotation axis is slightly out of the plane of the sky. That is $i = 76°$.

Similar models have been proposed by Radhakrishnan (Radhakrishnan and Cooke 1969) for the Vela pulsar. He also finds that the polarization axis is approximately perpendicular to the rotation axis. Erika Bohm-Vitense (1969) proposed an essentially similar model for the Crab Pulsar based on the shape of the optical light curve. Because the main pulse and the interpulse don't occur symmetrically our model would imply that the polarization axis is bent with both poles lying in the equatorial plane.

There's a striking similarity between these models and models for magnetic A stars. When one has good data for a magnetic A star one almost invariably finds that the magnetic axis is perpendicular to the rotation axis. The mass of the magnetic A star is of the order of 2 or 3 times the solar mass. This corresponds well to estimates of the mass of the Crab Nebula plus the estimated mass for the pulsar. The magentic field in magnetic A stars is higher than one needs to make a pulsar since only about 100 gauss (before collapse) is needed. For most main sequence stars with masses of about 2 or 3 solar masses one finds that a lot of angular momentum must be removed in order to make it collapse into an object 10 kilometers or so in diameter. With magnetic stars this may not be a problem. Magnetic stars have been found with rotation periods ranging from a fraction of a day, which would be too high to make a pulsar, to as long as 20 years (Preston, private communication). Preston thinks that there may be a strong interaction between the magnetic field of the star and the interstellar medium. If this is true it would mean that the old peculiar A stars are rotating slower than the younger ones and could presumably collapse to the pulsar configuration with no problem. The younger stars may have sufficient time for the interaction between the magnetic field and the interstellar medium to slow their rotation before they exhaust their nuclear fuel.

There may be a natural reason for the magnetic axis to be perpendicular to the rotation axis. Ostriker and Gunn (1969) conclude that the interaction between the magnetic field and the interstellar medium produces no torque to change the orientation of the spin axis. However, Mestel (1968) in analyzing a stellar wind model, concluded that an inclined magnetic axis would produce torques that would disappear for two orientations, the magnetic axis either parallel to the rotation axis or perpendicular to the rotation axis. Mestel didn't know which, if either, configuration was stable. The Crab pulsar and magnetic A stars seem to indicate that it's the perpendicular condition which is preferred by nature.

Now, what sort of problems does an optical astronomer have with pulsars? Well, we just don't have enough data. Only one optical object has been identified and there are some indications that it might be a peculiar object. It may be true that the simple model we have for the Crab Nebula pulsar will not work for all the radio observations. The often complicated structure seen in the radio polarization data indicates that perhaps this is too simple a model. We wish we could see some more pulsars at optical wavelengths.

Of course the main theoretical problem is that we do not know the microscopic physics of the light-producing source.

Where do we go from here? One thing that needs to be done is to measure the rate of change of the period and the second derivative. The first derivative is known for the Crab. That gives you the net energy that the pulsar is pumping into the Crab Nebula. Various theories exist for second derivatives, and two groups, Princeton and Arecibo, have measured the second derivative with very small uncertainties but obtained values different by a large factor. There is at least one object, the pulsar in Vela, in which the sign of the first derivative changed for at least a short period of time.

Another question to be answered is whether there is substructure in the light variation of the pulsar. At the present time the available evidence suggests that such substructure is either absent or weak. The analysis of data that has already been taken should allow the present upper limits to be lowered.

Work with X-rays will help us with the far ultraviolet spectrum, and continued polarization studies may show precession of the spin axis of the pulsar.

DISCUSSION

LENCHEK: "Can you see any relation between the configuration of the pulsar and the light ripples?"

WAMPLER: "Jeff Scargle has made photographs of the ripples. They seem to be aligned along a certain axis which lies just about in the position we get for the rotation axis of the star. Whether that's significant or not I don't know, but it seems to be a fact."

QUESTION: "Do you have any idea how the magnetic axis of the Crab is aligned relative to the general magnetic field of the Crab?"

WAMPLER: "Well, for one thing, the magentic or polarization axis of the pulsar is perpendicular to the rotation axis. And the magnetic field of the Nebula is not all that regular, so the answer depends upon what you mean by the magnetic field of the Nebula."

G. WESTERHOUT: "But the rotation axis seems to be aligned parallel to the general field of the Nebula."

WAMPLER: "It is parallel to the magnetic field in the nearby nebulosity."

REFERENCES

Bohm-Vitense, E. 1969, Ap. J. (Letters), 156, L131.

Bradt, H., Rappaport, S., Mayer, W., Nather, R.E., Warner, B., MacFarlane, M., and Kristian, J. 1969, Nature, 222, 728.

Cocke, W.J., Disney, M.J., and Taylor, D.J. 1969, Nature, 221, 525.

Conklin, E.K., Howard, H.T., Miller, J.S., and Wampler, E.J. 1969, Nature, 222, 552.

Duthie, J.G., Sturch, C., Richer, H.B., and Rodnay, P. 1969, Science, 163, 1320.

Freeman, K.C., Rodgers, A.W., Rudge, P.T., and Lynga, G. 1969, Nature, 222, 459.

Goldstein, S.J., and Meisel, D.D. 1969, Nature (in press).

Kristian, J. 1969, Ap. J. (Letters), 154, L99.

_____ 1969, Pub. A.S.P. (in press).

Lynds, R., Maran, S.P., and Trumbo, D.E. 1969, Ap. J. (Letters), 155, L121.

Mestel, L. 1969, M.N., 140, 177.

Miller, J.S., and Wampler, E.J. 1969, Nature, 222, 552.

Minkowski, R. 1942, Ap. J., 96, 199.

Neugebauer, G., Becklin, E.E., Kristian, J., Leighton, R.B., Snellen, G., and Westphal, J.A. 1969, Ap. J. (Letters), 156, L115.

Oke, J.B. 1969, Ap. J. (Letters), 156, L49.

Ostriker, J.P., and Gunn, J.E. 1969, Ap. J., 157, 1395.

Radharkrishnan, V., and Cooke, D.J. 1969, Ap. Letters, 3, 225.

Scargle, J.D. 1969, Ap. Letters, 3, 73.

Wampler, E.J., Scargle, J.D., and Miller, J.S. 1969, Ap. J. (Letters), 157, L1.

Warner, B., Nather, R.E., and MacFarlane, M. 1969, Nature, 222, 233.

September 29, 1969

X-RAY OBSERVATIONS OF PULSARS*

Hale V. Bradt

Center for Space Research
Massachusetts Institute of Technology
Cambridge, Mass., U.S.A. 02139

Observations of pulsars at X-ray wavelengths must be made from above the atmosphere. Accordingly, the observations have been fewer in number and less comprehensive than those at optical and radio wavelengths. The X-ray pulsations from NP 0532, discovered by the NRL group (Fritz et al., 1969) (Fig. 1), have been studied by groups at Massachusetts Institute of Technology (Bradt et al., 1969; Floyd, Glass and Schnopper, 1969), at Goddard Space Flight Center (Boldt et al., 1969), and Rice University (Fishman, Harnden, and Haymes, 1969; Fishman et al., 1969). In each case the detectors had wide fields of view and the entire Crab Nebula-NP 0532 radiation was detected. Thus, for instance one cannot exclude a continuous, non-pulsing component of X-rays from the pulsar itself. Earlier angular size measurements (Bowyer et al., 1964; Oda et al., 1967) indicate that the bulk of the Crab radiation comes from a region about 100" in diameter. However, even these measurements cannot exclude a non-pulsing component equal in magnitude to the pulsing component.

The published data obtained from sounding rockets (Fritz et al., 1969; Bradt et al., 1969; Boldt et al., 1969) at 1-10 keV yield the following features of the X-ray pulsations (Fig. 2b) (1) The ratio of pulsed to unpulsed radiation is about 10%. There is some discrepency in the reported percentages which might indicate time variations in intensity over a period of about 1 year (Boldt et al., 1969); (2) the two pulses are about equal in energy flow whereas, in the optical, the secondary peak is significantly reduced compared to the primary; (3) the primary peak is considerably narrower than the secondary peak and both widths are comparable to the optical widths; (4) the fractional background between pulses is considerably greater in the X-ray than in the optical; (5) the rise and fall times of the secondary pulse differ from the optical.

The relative arrival times of the optical and X-ray pulses were obtained from nearly simultaneous observations by the groups at MIT,

*The MIT work described herein was supported in part by NASA under Contract NSR 22-009-129 and grant NGR 22-009-015. This talk was presented in essentially the same form at the Symposium on Pulsars, Rome in December 1969 and will appear in the proceedings of the symposium.

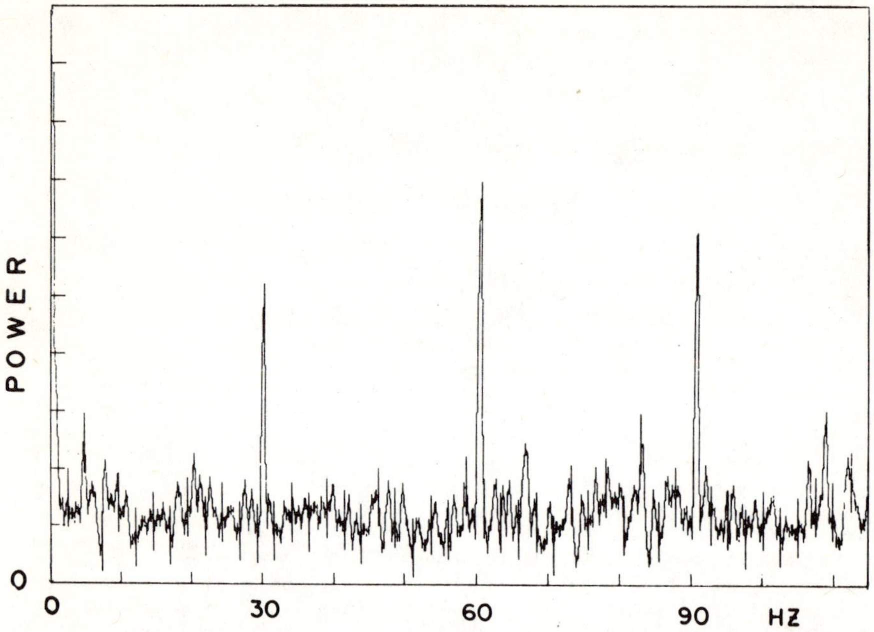

Fig. 1. Power density spectrum from NRL data of X-radiation from the Crab Nubula. The peaks represent NP 0532. This result gave the first indication that NP 0532 is pulsing at X-ray wavelengths. (Fritz et al., 1969)

Mt. Palomar Observatory, and McDonald Observatory (Bradt et al. 1969). The principal optical peak was found to coincide, to within 1 msec, with the narrower X-ray peak (Fig. 2).

At balloon altitudes, two groups (Floyd, Glass, and Schnopper, 1969; Fishman et al., 1969a and 1969b) obtained data which clearly indicate that about 15% of the radiation from the Crab at energies above 30 keV is pulsing (Fig. 3). The Rice University observations were made on June 4, 1967. The authors noted that the period of the pulsar obtained from this data differed by 16 ± 6 nsec from that obtained from an extrapolation to 1967 of the recent radio values of dP/dt and d^2P/dt^2. It has been pointed out by Ramaty and Holt (1969) that this discrepancy, if real, is consistent with discontinuous speed ups of the pulsar on the order of $\Delta T \simeq 1$ nsec every 100 days. This is much larger than the observed speedup of 8 x 10^{-11} sec on Sept. 28, 1969.

The percentages quoted above indicate that the spectrum of the pulsar in the X-ray region is nearly parallel to that for the Nebula as a whole (spectral index ∿1.0), though there is some indication that the pulsar spectrum is slightly flatter. In contrast, a comparison of the spectral points in the optical region (Oke, 1969) to that at 3 keV yields a much flatter spectrum (index 0.5) for the pulsation than for the continuous radiation. The consequence of this is that the ratio of energy flux in X-rays relative to the optical is:

$I_x/I_o \sim 100$ NP 0532

$I_x/I_o \sim 2$ Crab

Comparison to the radio flux from NP 0532 yields:

$I_x/I_R \sim 3 \times 10^5$ NP 0532

The energy flux in the 1.5-10 keV interval is:

$$I_x = 2.2 \times 10^{-9} \text{ ergs cm}^{-2} \text{ sec}^{-1}$$

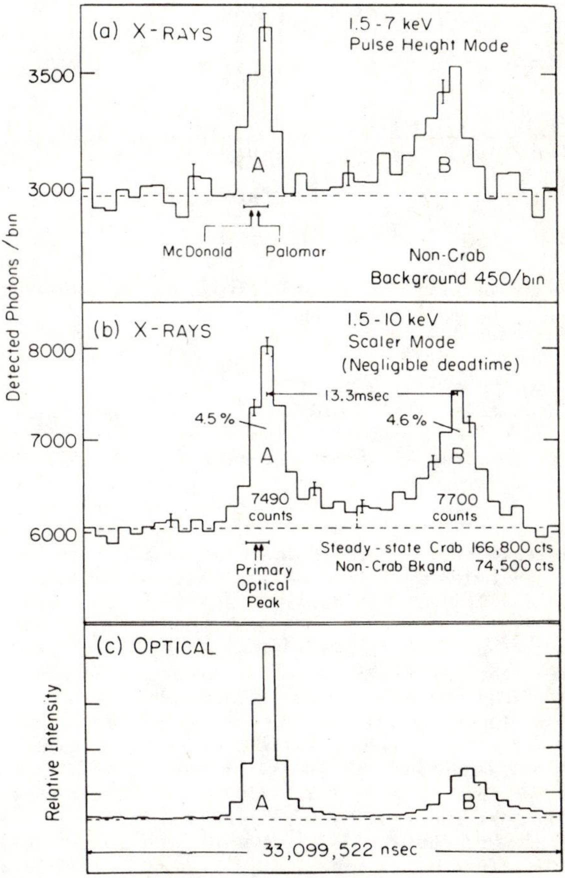

Fig. 2. Previously reported shapes of pulses from NP 0532 and comparison to optical phase. Upper histograms was the first indication to MIT groups of X-ray pulsing. The shape is best obtained from the middle histogram. (Bradt et al., 1969)

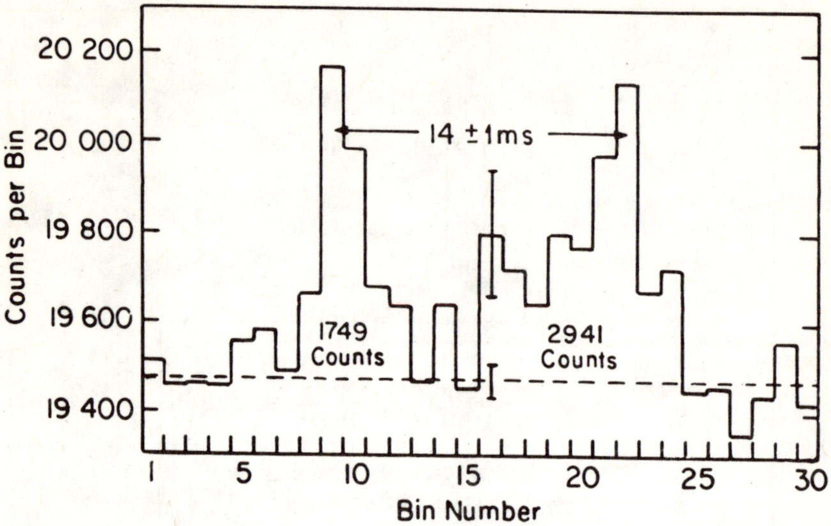

Fig. 3. Shape of NP 0532 at 25-100 keV from balloon altitudes. (Floyd, Glass, and Schnopper, 1969)

Recent reanalyses of the MIT X-ray data (Rappaport, Bradt, and Mayer, 1970) from 1.5-10 keV yield the following results. The final time standard, with an uncertainty of ±0.1 msec was used to superimpose all of the data with the apparent periodicity of NP 0532 at White Sands Missile Range. This superimposition (Fig. 4), with about twice the quantity of data previously reported, yields no substantial shape of the average X-ray pulse shapes. However, the improved timing, together with signal processing delays, indicates a coincidence of optical and X-ray pulses within about 0.3 msec (Fig. 4). Comparison with the optical data of the Harvard and Princeton observatories might yield a somewhat better limit.

A superposition of the data in 200 μsec bins was carried out to study the pulse shape of the principal X-ray peak. A width of 1600±100 μsec (FWHM) was obtained. This is substantially greater than the 1000-μsec timing resolution of our signal processing system. This indicates an intrinsic X-ray pulse width of about 1.2±0.2 msec which is to be compared to an optical width of 1.4 msec. The NRL data have intrinsically much better timing resolution. They report (Fritz et al., 1970) (Fig. 5), that their analysis at present indicates a primary pulse width of 1.0 msec and that final calibrations of their time standard could reduce this further. At present, it is not yet apparent whether or not there is a significant difference between the optical and X-ray pulse widths. Further analysis of both experiments would clarify this point.

We have searched for time variations in the X-ray intensity of NP 0532 during the course of the 3 min observations with negative

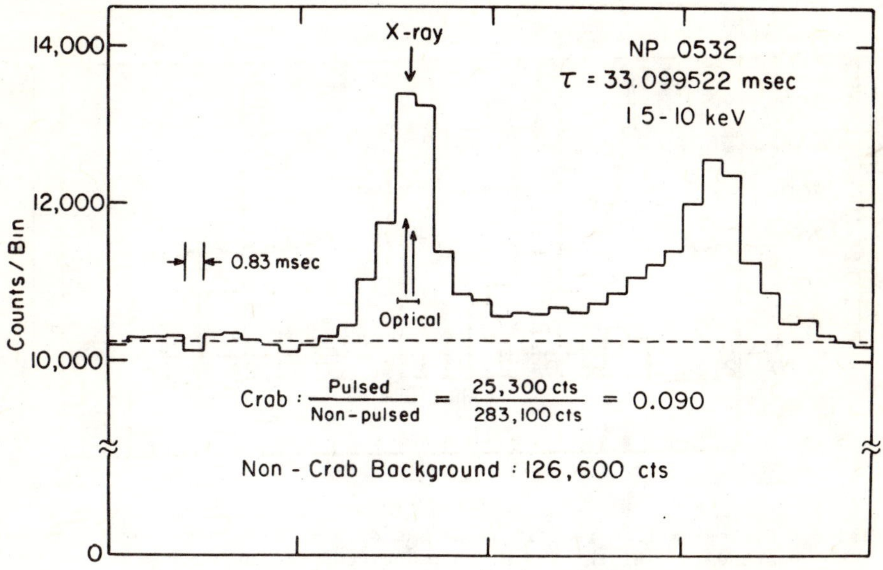

Fig. 4. Recent superposition of MIT data 1.5-10 keV with about twice the data and with an improved timing reference for the X-ray data. The optical peak times (Bradt et al., 1969) are from McDonald and Mt. Palomar Observatories. (Rappaport, Bradt, and Mayer, 1970)

results: (1) Fourier analyses of the data yield no other frequencies above background; (2) Superpositions with twice the pulsar period indicate no significant modification of pulse shape from odd to even pulses (Fig. 6); (3) Superpositions of data from successive 20 sec intervals indicate no gross time variations (Fig. 7).

We have also searched for low energy radiation from NP 0532. Aluminum-window counters on the MIT rocket provided a well defined spectral window from 1.0-1.5 keV. The results of a superposition (Fig. 8) of these data indicate that 2.5±0.9% of the total radiation in this energy band is pulsed and lies in the primary pulse. The secondary peak is not apparent above the background. Thus at 10Å, pulsed radiation is still observed but apparently at about 1/2 the fractional intensity observed at 4Å.

Searches for X-ray pulsations from other X-ray sources have been carried out by the NRL (Friedman et al., 1969), AS&E (Gorenstein et al., 1970b) and MIT groups. The summarized results are given in Table I. The upper limits of the percentages of energy in any frequency component relative to the total X-ray energy are given. Note that the limit depends on the type of analysis and assumed pulse shapes used by the authors. It should be noted that two other supernova remnants (Cas A, Tycho) are included in Table I. A lack of statistics precluded a lower threshold in the search. Also note that the AS&E search has been carried out only for the range of periods 8 to 35 msec.

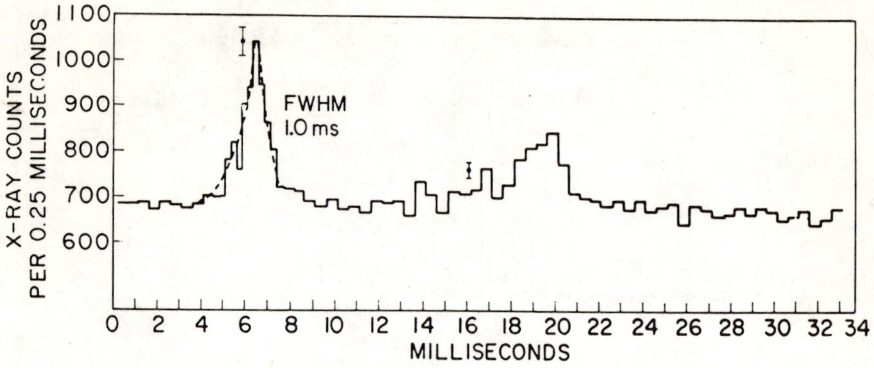

Fig. 5. Histogram from reanalysis of NRL data with improved time standard. Width of peak is to be compared to optical width of 1.4 msec. (Fritz et al., 1970)

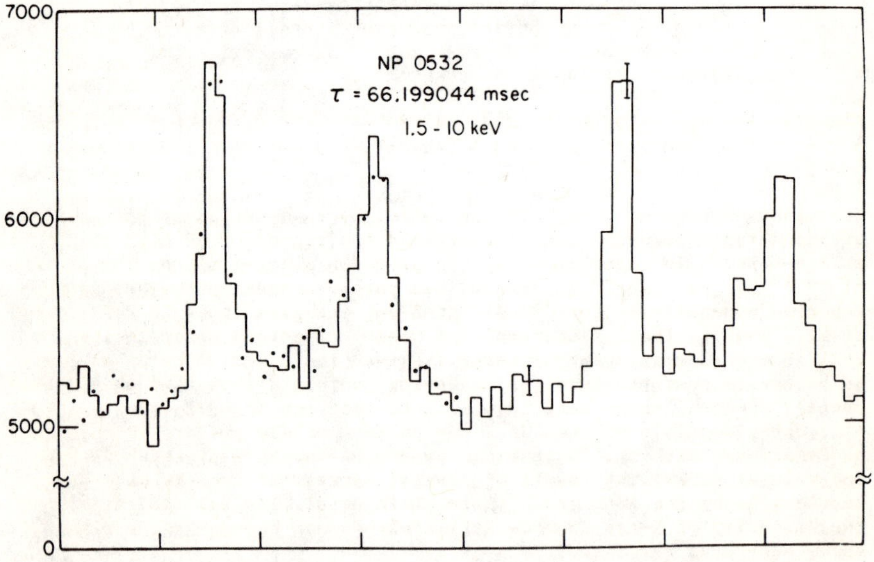

Fig. 6. Superposition of MIT data with double period. For comparison, the right side of the histogram is reproduced on the left side in the form of dots. (Rappaport, Bradt, and Mayer, 1970)

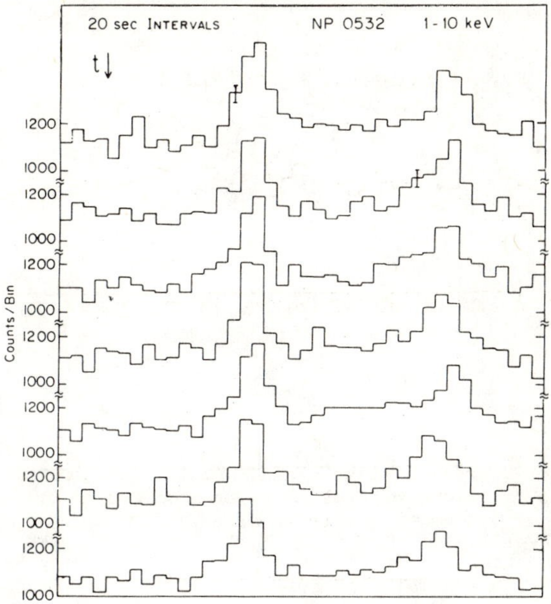

Fig. 7. The MIT data for NP 0532 broken into seven 20-sec intervals. (Rappaport, Bradt, and Mayer, 1970)

Another type of search is for pulsed X-ray emission at the locations of known radio pulsars. We have studied the region of CP 1133 and NP 0527 at 1.5-6 keV. In either case we can place an upper limit on the order of 10^{-10} erg cm^{-2} sec^{-1} for the pulsed power. Recently the UCSD group reported upper limits for CP 1919 at X-ray energies above 20 keV (Gruber et al., 1969). Their summary of the observations and upper limits for CP 1919 over a wide spectral range is given in Fig. 9. The upper limit at 50 keV is just the level of intensity one would expect if the CP 1919 overall spectral shape were identical to that for NP 0532.

Other possible sources of X-ray pulsations are pulsars which might be associated with the 25 probable supernova remnants listed by Poveda and Woltjer (1969). A result of several surveys of the sky at X-ray wavelengths by the AS&E group (Gorenstein et al., 1970a) indicates that the "majority of X-ray sources definitely do not belong to the class of known supernova remnants."

Finally, I would like to comment on the recent suggestion by Slysh (1969) that all of the X-rays from the Crab Nebula are emitted by the pulsar in the pulsating mode. He points out that scattering of about 90% of these by interstellar grains would account both for the non-pulsed character of the radiation as well as the apparent finite angular size of the source. The latter point was first suggested by Overbeck in

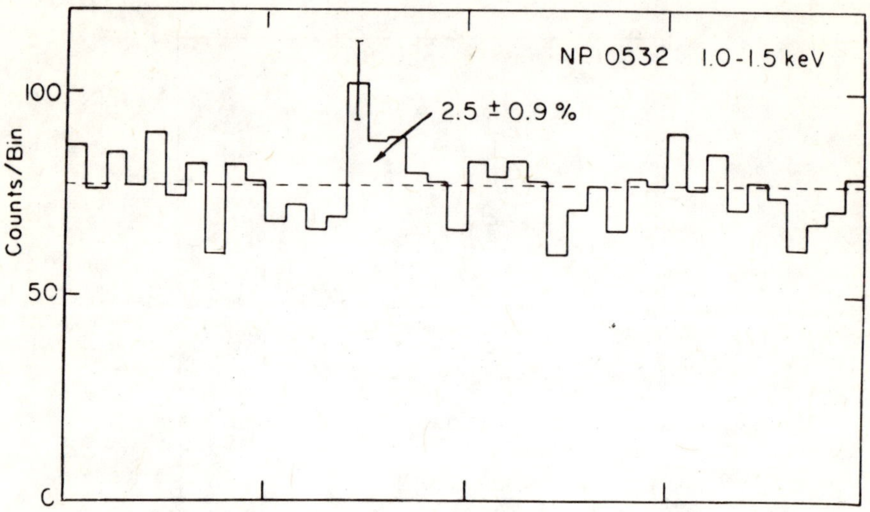

Fig. 8. Superposition of NP 0532 X-ray data at 1.0-1.5 keV (∿10Å). The spectral window is defined without ambiguity by the transmission characteristics of the aluminum windows on the proportional counters. The percentage is the ratio of pulsed to unpulsed radiations from the Crab Nebula. (Rappaport, Bradt, and Mayer, 1970)

TABLE I

UPPER LIMITS ON PULSATIONS FROM X-RAY SOURCES

Source	Photons	Periods(sec)	Limit	Group
Sco X-1	600,000	.020-20	0.5%	MIT[a]
GX5-1	21,000	.010-15	2%	"
GX9+1	5,800	.010-7	4%	"
GX13+1	4,000	.007-4	5%	"
GX349+2	2,400	.007-2	6%	"
Sco X-1	64,000	.070-4	1%	NRL[b]
Cyg XR-1	3,200	.008-4	6%	"
Cyg XR-2	3,200	.008-4	6%	"
Cas A	900	.008-.035	15%	AS&E[c]
Tycho SN	250	.008-.035	20%	"

a. (Rappaport, Bradt, and Mayer, 1970)
b. (Friedman et al., 1969)
c. (Gorenstein et al., 1970b)

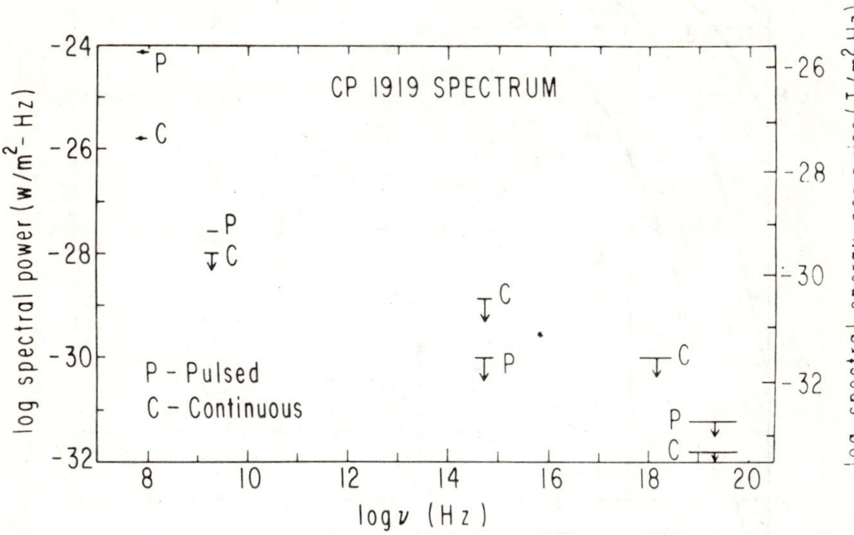

Fig. 9. Spectral power versus frequency for the continuous (C) and pulsed (P) radiation from CP 1919. (University of California, San Diego, Gruber et al., 1969)

the context of the early lunar occultation experiment of NRL which was designed to test the hypothesis that a neutron star was the source of the X-radiation from the Crab Nebula. A calculation of the effect of such scattering on the angular size of an X-ray source was carried out by Overbeck (1965). An immediate consequence of his analysis (Fig. 10) was that, while such scattering is potentially important for X-rays emitted from the Crab Nebula, the presently accepted values of the sizes and numbers of grains (e.g., Allen, 1963) would cause only a small fraction (∿5-10%) of the radiation at 3 keV to be scattered into angles on the order of only 20". His results (Fig. 10) include the effect of double scatters and the dominant effect of the larger grains which yield a greater scattering probability and a smaller average scattering angle than might otherwise be expected. Naranan and Shah (1969), Bowyer, Lampton, and Mack (private communication), and Ryter (1970) have studied in depth the available evidence on reddening and conclude in independent calculations and considerations that it is highly improbable that the scattering by grains would be a significant effect at 3 keV.

In the pulsar data, the fraction of X-rays which are pulsed does not vary by more than a factor of 3 over an energy range of 100 (1-100keV). The scattering hypothesis predicts an effect which varies as λ^2. Of course peculiar emission spectral shapes for the nebular and pulsar X-ray radiations could conspire to mask this effect. Thus the definitive test of the Slysh hypothesis appears to be the measurement of angular size as a function of X-ray energy. Preliminary results (Floyd, private communication) from a balloon flight by an MIT group indicate that the X-ray size of the Crab Nebula at 30 keV is greater than ∿1 arc minute

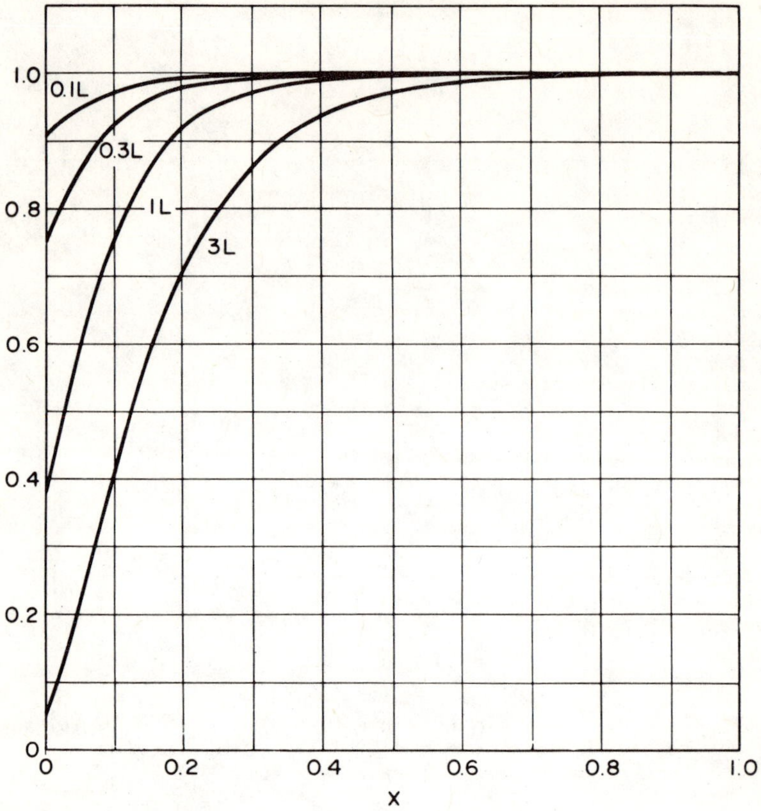

Fig. 10. The fraction of X-rays received within an angle $x\lambda/a_o$ from the direction of a point source as a function of L, the mean free path for scattering (from Overbeck, 1965). The number of mean free paths to NP 0532 is about 0.1 at λ = 3Å if we consider ice grains with a typical density of 2 x 10^{-13} grains/cm^3 (Allen, 1963) and a Gaussian distribution of radii with average radius a_o = 0.04μ. The characteristic scattering angle in this case is λ/a_o = 3.4'. However, note that the actual scattering angles are less than 10% of this value. The same grains yield 1.5 mean free paths at 10Å (∿30% of X-rays are unscattered). The recent work of Naranan and Shah (1969) yields even smaller effects, i.e., ∿0.15 mean free paths at 10Å.

whereas the scattering hypothesis clearly requires a point source at this energy. Further measurements of the angular structure at these energies as well as at 10-44Å are clearly of importance since they will give insight into the nature of the X-radiation from the Crab Nebula and NP 0532 as well as the character of the interstellar grains.

In summary, I will reiterate that the X-radiation from NP 0532 represents by far the greatest emitted power among the several observable wavelength regions. Thus to gain insight into the nature of the emission processes it is clearly important to pursue studies of the character of this radiation, i.e. pulse shapes, time variations, the angular structure of the Nebula-pulser X-ray emission, etc. Also, a continuing search for other X-ray pulsars is important because it is possible that such objects would not radiate in the radio and optical regions at detectable levels.

REFERENCES

Allen, C.W., Astrophysical Quantities (Oxford University Press, New York, 1963) 2nd ed.

Boldt, E.A., Desai, U.D., Holt, S.S., Serlemitsos and Silverberg, R.F. 1969, Nature, 223, 280.

Bowyer, S., Byram, E.T., Chubb, T.A., and Friedman, H. 1964, Science, 146, 912.

Bowyer, S., Lampton, M., and Mack, J., private communication.

Bradt, H., Rappaport, S., Mayer, W., Nather, R.E., Warner, B., MacFarlane, M., and Kristian, J. 1969, Nature 222, 728.

Fishman, G.J., Harnden, F.R., Jr., and Haymes, R.C. 1969, Ap. J. (Letters) 156, L107.

Fishman, G.J., Harnden, R.R., Jr., Johnson, W.N., III., and Haymes, R.C. 1969, Ap. J., 158, L61.

Floyd, F., private communication.

Floyd, F.W., Glass, I.S., and Schnopper, H.W. 1969, Nature, 224, 50.

Fritz, G., Henry, R.C., Meekins, J.F., Chubb, T.A., and Friedman, H. 1969, Science, 164, 709.

Fritz, G., Henry, R.C., Meekins, J.F., Chubb, T.A., and Friedman, H. 1970, in preparation.

Friedman, H., Fritz, G., Henry, R.C., Hollinger, J.P., Meekins, J.F., and Sadeh, D. 1969, Nature, 221, 345.

Gorenstein, P., Gursky, H., Kellogg, E., and Giacconi, R. 1970, Ap. J., in press.

Gorenstein, P., Kellogg, F.M., and Gursky, H. 1970, Ap. J., in press.

Gruber, D., Matteson, J.L., Pelling, M.R., and Peterson, L.E. 1969, submitted to Nature.

Naranan, S., and Shah, G.A. 1969, submitted to Nature.

Oda, M., Bradt, H., Garmire, G., Spada, G., Sreekantan, B.V., Gursky, H., Giacconi, R., Gorenstein, P., and Waters, J.R. 1967, Ap. J., 148, L5.

Oke, J.B. 1969, Ap. J. (Letters), 156, L49.

Overbeck, J., 1965, Ap. J., 141, 864.

Poveda, A., and Woltjer, L. 1968, Astron. J., 73, 65.

Ramaty, R., and Holt, S.S. 1969, Nature, 224, 1003.

Rappaport, S., Bradt, H., and Mayer, W. 1970, in preparation.

Ryter, C. 1970, preprint.

Slysh, V.I. 1969, Nature, 224, 159.

October 20, 1969

THE MEASUREMENT OF PULSAR PERIODS

G.S. Downs

Jet Propulsion Laboratory
California

In about the middle of 1968 G.A. Morris, Paul Reichley and I at Jet Propulsion Laboratory began weekly observations of CP 0950. The hope at that time was that by tracking 0950 for a year one could analyze the time of arrival data and determine both position and distance. It now seems that the distances are too great to be resolved and instead we have to be happy with obtaining right ascension, declination, the period and its first and second derivatives. As time progressed we added eleven more pulsars to our program and at this time we are now tracking twelve, each at least once a week. We are using an 85 ft. antenna of the Goldstone Tracking Station at a frequency of 2388 MHz. The 20° system temperature makes this project possible.

The times of arrival are measured with standard deviations as low as 30 μs. A cesium standard provides our time base. The ephemeris used in correcting arrival times to the barycenter was the Jet Propulsion Laboratory's DE-43. The sources are considerably weaker at 12.5 centimeters, with spectra that are falling off with an index of less than -2 in most cases, so we have to integrate over a number of pulse intervals before we obtain decent signal-to-noise ratios. We therefore integrate over 500 pulses. In order to superimpose sucessive groups of 500 pulses we must employ a frequency synthesizer that is matched to the frequency of the pulsar. If the frequency synthesizer is not well matched to the pulsar frequency, the pulse will be smeared out by the addition of successive groups of pulses and you will never see the pulsar (Downs, Morris, and Reichley, 1969). Because the pulses have a predominantly triangular shape, we run a triangular correlation with the output. The maximum value of this correlation determines the arrival time of the integrated pulses relative to the integration start-time. Some sample pulse shapes are shown in Figures 1 and 2. The pulses are made up of one to three components (for example, PSR 0329 + 54 and PSR 2045 - 16), a fact neglected in current theoretical treatments.

The problem of an incorrect synthesizer setting plagued us during February 1969 observing PSR 0833-45 (the pulsar in Vela). We had noted that the period was changing at something like 10 nanoseconds a day, a result that was published by the Australian observers at Parkes. Each time we sent out period predictions to the operators at Goldstone we had to use an up-to-date period. We had slipped a day late in January and went for about a month with the integrated pulses smearing slightly. We caught up with that mistake late in February, but a week later the pulses were smearing again. This smearing went on for about a month. We couldn't explain it as an experimental error, so we concluded that something had happened to the pulsar. We asked the Australians if they had also seen an abrupt period change. We sent a telegram to Radhakrishnan and he quickly returned a positive reply by telephone. In our mutual amazement we got together and rapidly published the result that the period had in fact decreased abruptly (Radhakrishnan and Manchester, 1969; Reichley and Downs, 1969).

Let me discuss now how we measure the period. In order to compute the time of arrival of a pulse we begin by calculating the right ascension and declination of the earth relative to the barycenter, the center of mass of the solar system. We then use that position and the position of the pulsar to calculate what the time of arrival would have been had our antenna been at the barycenter. The varying speed of the earth and its immersion in a varying gravitational field introduce a relativistic clock correction which amounts to as much as 1.7 milliseconds (Melbourne et al., 1968; Clemence and Szebehely, 1967).

We write the time of arrival of the n-th pulse t_n at our antenna as

$$t_n = t_{ne} + D(t_{ne})/c - t_o \qquad (1)$$

where t_{ne} is the time of emission at the source, D/c is the flight time an and t_o is a reference time. The objective is to solve for a model describing t_{ne}. The correction for the variation in our clock in orbit around the sun is

$$\Delta t_n = A(\cos\delta_e \cos\alpha_e - \cos\delta_{oe} \cos\alpha_{oe}) \qquad (2)$$

where $A = 1.66 \times 10^{-3}$ seconds, and $(\alpha_{oe},\delta_{oe})$ is the position of the earth at t_o.

We assume we can represent the pulse period by a Taylor expansion,

$$P(t) = P_o + P_1 t + P_2 t^2, \qquad (3)$$

and we represent the emission time by

$$t_{ne} - t_o = c_1 n + c_2 n^2 + c_3 n^3 \qquad (4)$$

where n is the number of pulses since t_o and the c_i'are functions of P_i.

The pulse number is kept track of by our approximate knowledge of the period. Of course, the uncertainty in the period defines how long we can observe before we lose track of which pulse is arriving. The term $\cos\delta_{oe} \cos\alpha_{oe}$ in (2) is dropped for this illustration. If the pulsar is not accelerating relative to the barycenter then the term D/c in (1) can be absorbed into the reference time. We then have, assuming the Earth-barycenter distance $r \ll D$,

$$t_n - (c_1 n + c_2 n^2 + c_3 n^3) + A \cos\delta_e \cos\alpha_e$$

$$= \frac{r}{c} [\cos (\alpha_p - \alpha_e) \cos\delta_p \cos\delta_e + \sin\delta_e \sin\delta_p] \qquad (5)$$

$$= \frac{1}{c} \times$$ (the distance from the earth, in the direction of the pulsar, to the plane containing the barycenter and normal to the earth-pulsar line).

Here α and δ are right ascension and declination, respectively. The subscripts p and e refer to the pulsar and the earth, respectively, and r is the Earth-barycenter distance.

We have one equation like (5) for each measurement of pulse arrival time. Any solution for the coefficients c_i will be polluted by errors in the pulsar position. We are therefore forced to solve for all five

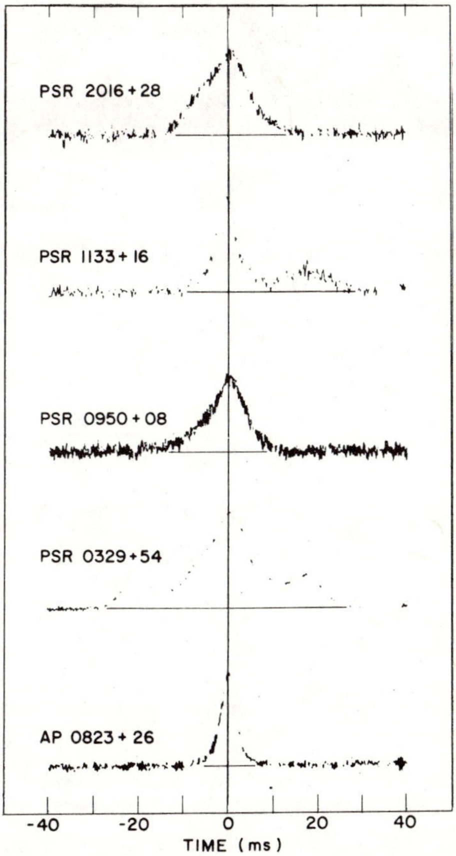

Fig. 1. Pulse shapes of several sources. Note the predominantly triangular shapes.

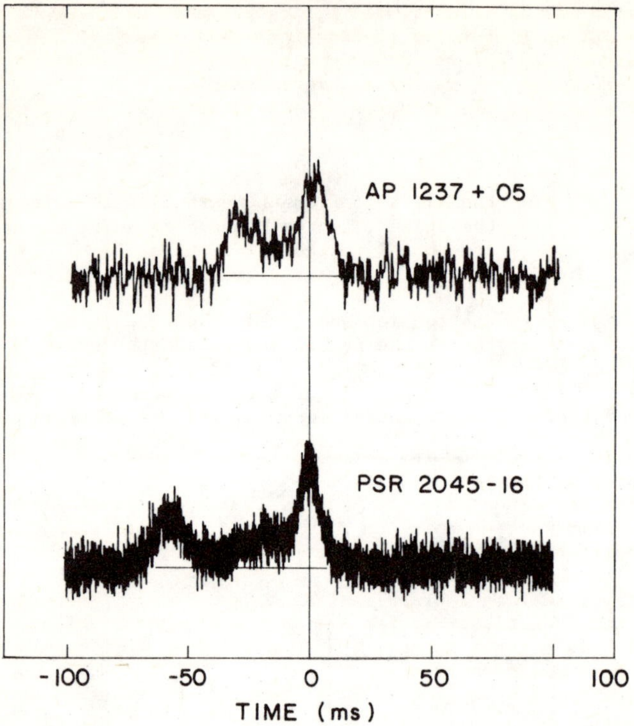

Fig. 2. Pulse shapes. Note the three components in PSR 2045-16.

parameters c_1, c_2, c_3, α_p and δ_p simultaneously. Fortunately we begin with a fair knowledge of these parameters so we replace α_p with α_{op} + $\Delta\alpha_p$, etc., thereby linearizing the problem. The time-of-arrival residual ΔT_n is then written as

$$\Delta T_n = \text{measured minus predicted arrival time}$$

$$\Delta T_n = t_n - (c_{10}n + c_{20}n^2 + c_{30}n^3) + A \cos\delta_e \cos\alpha_e + \frac{r}{c} f\,(\alpha_{po}, \delta_{po}, \alpha_e, \delta_e)$$

$$\Delta T_n = n\,\Delta c_1 + n^2\,\Delta c_2 + n^3\,\Delta c_3 + [\cos\delta_{po} \cos\delta_o \sin(\alpha_e - \alpha_{po})]\,\Delta\alpha_p + [\sin\delta_e \cos\delta_{po} - \cos\delta_e \sin\delta_{po} \cos(\alpha_e - \alpha_{po})]\,\Delta\delta_p . \quad (6)$$

Now, $\Delta c_1 = \Delta p_o$ so an error in the period produces a linear deviation in t_n. $\Delta c_2 = \Delta p_1^o$ so a changing period produces a quadratic deviation. Position errors $\Delta\alpha_p$ and $\Delta\delta_p$ produce a sinusoidal variation in the difference between t_n (measured) and t_{ne} (predicted).

These equations are arranged in matrix form for the series observations as

$$\begin{pmatrix} \Delta T_i \\ \cdot \\ \cdot \\ \cdot \\ \cdot \end{pmatrix} = \begin{pmatrix} n_i & n_i^2 & n_i^3 & \cdots \\ & \cdot & & \\ & \cdot & & \\ & \cdot & & \\ & \cdot & & \end{pmatrix} \begin{pmatrix} \Delta c_1 \\ \Delta c_2 \\ \Delta c_3 \\ \Delta\alpha_p \\ \Delta\delta_p \end{pmatrix}$$

and solved in a least-squares sense for Δc_1, Δc_2, Δc_3, $\Delta\alpha_p$ and $\Delta\delta_p$ (see e.g., Chapter 4, Hamilton, 1964).

So with the crude model of the position and a rough idea of the period an analysis like this can give period, rate of change of period and the position. The data on a number of pulsars is collected in Table 1. The most exciting numbers of this table are the second derivatives. If you replace the Arecibo (Richards, Rankin, and Counselman, 1969) value for the Crab listed in Table 1 by the Princeton (Boynton et al., 1969) value, which is 4 times larger, then you see that the Crab and Vela have about the same second derivatives.

In Figure 3 I have plotted $\dot{P}$ vs. the period. There's considerable scatter about a straight line. If we write $\dot{P} \propto P^\beta$ then the various theories for the decay of a pulsar predict different values of β. Figure 3 shows the wide variation in the value of β that can be fit to the data. I call your attention to the following anomaly. PSR 2045 has the third highest value of $\dot{P}$ but the longest period in this group, while PSR 0950 + 08 has a very low value of $\dot{P}$ and yet it is one of the faster pulsars. This seems to point to a wide range of physical conditions in the pulsars.

We have several sources under observation now for which we don't yet have values for the rate of change of period. We have to observe a source for several months before we can separate an effect due to $\dot{P}$ from those due to position errors. These sources are indicated in Table 1 by brackets in the column headed $\dot{P}$.

The pulsar positions that we have obtained from our analysis are shown in Table 2. These positions are in good agreement with positions obtained by other observers. PSR 2045 shows how the position of this source relative to the ecliptic affects our analysis. This source has a very low celestial latitude and as a result we get a very poor determination of the position even after 7 or 8 months of observing.

Applying the above technique to the measurement of the period of PSR 0833-45 we find that when the period jumped at the end of February 1969; the difference between the extrapolated and measured period is 208 ns. This is close to the value found by Radhakrishnan and Manchester. This amounts to a fractional change in period of 2×10^{-6}. After the jump the rate of change of period ($\dot{P}$) was about 1% higher than it was before the jump (see Figure 4).

We have reconstructed the behavior of PSR 0833-45 through the month of June 1969 and this is shown in Figure 5. In addition to the 1% change in $\dot{P}$, the curve of P(t) is now slowly curving over toward the slope that it had before March, 1969.

TABLE 1

PERIOD PARAMETERS

PULSAR	P(sec)	$\dot{P}$(sec/sec) x 10^{15}	P(Sec/sec^2)	Obs.
PSR 0329 + 54	0.71451	1.9 ± 0.2		Camb.
		2.06 ± 0.01		JPL
PSR 0531 + 21	0.03310	422.66 ± 0.02	(-3.2 0.08) x 10^{-24}	AIO
CP 0808	1.29224	< 0.3		Camb.
AP 0823 + 16	0.53066	1.38 ± 0.01		JPL
PSR 0833 - 45	0.08921	125.0 ± 0.1	(-3.2 0.1) x 10^{-23}	JPL
		120		CSIRO
PSR 0834 + 06	1.27376	5.0 ± 0.8		Manch.
		6.3		Camb.
PSR 0950 + 08	0.25306	0.3 ± 0.1		Manch.
		0.229 ± 0.005		JPL
PSR 1133 + 16	1.18791	4.1 ± 0.5		Manch.
		3.2		Camb.
		3.78 ± 0.02		JPL
AP 1237 + 25	1.382	( )		JPL
PSR 1508 ± 55	0.73968	5.3 ± 0.4		Camb.
PSR 1642 - 03	0.3877	( )		JPL
PSR 1749 - 28	0.5625	( )		JPL
PSR 1919 + 21	1.33730	1.1 ± 0.5		Manch.
		0.95		Camb.
PSR 1929 + 10	0.227	( )		
PSR 1933 + 16	0.35873	6.01 ± 0.04		JPL
PSR 2016 + 28	0.55795	0.146 ± 0.010		JPL
PSR 2045 - 16	1.96157	11.3 ± 0.3		JPL

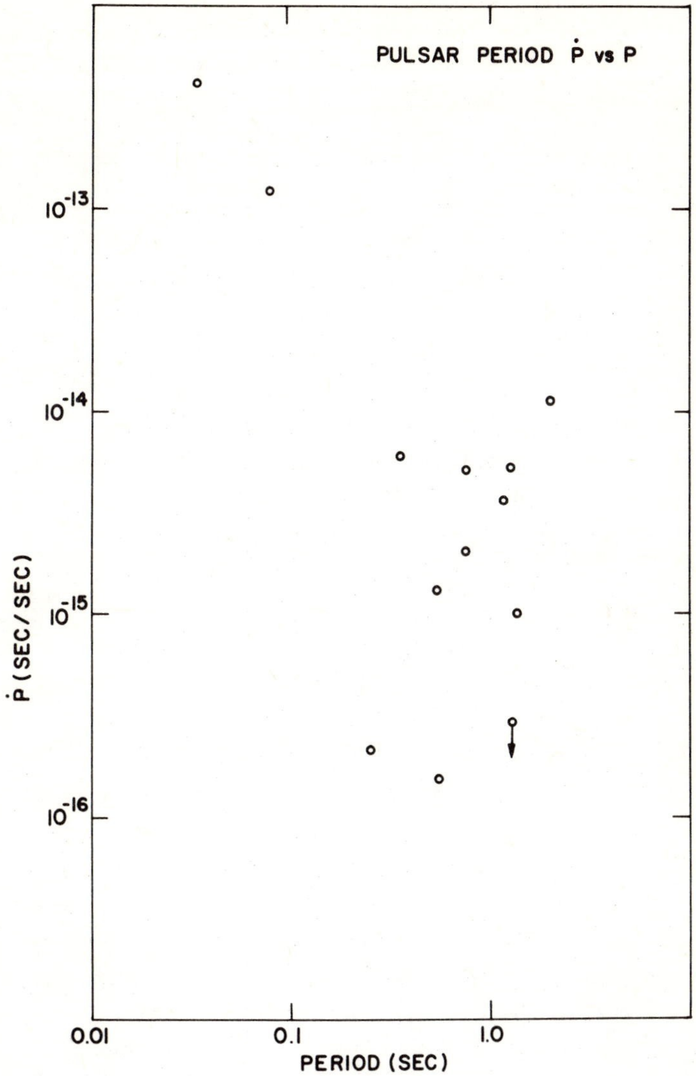

Fig. 3. Rate of change of period versus period for several sources.

TABLE 2

PULSAR POSITIONS

PULSAR	R.A. (1950.0)	Dec. (1950.0)	OBS.
PSR 0329 + 54	$03^{h},29^{m},11^{s}.1 \pm 0^{s}.1$	$54° \; 24' \; 37''.0 \pm 1''.5$	JPL
	$11^{s}.0$	$37''.2$	Manch.
AP 0823 + 26	$08^{h},23^{m},52^{s} \pm 1^{s}$	$26° \; 47' \; 21'' \pm 3''$	JPL
	$52^{s} \pm 10^{s}$		Meudon
		$48' \pm 5'$	AIO
PSR 0833 - 45	$08^{h},33^{m},39^{s}.01 \pm 0^{s}.03$	$-45° \; 00' \; 06'' \pm 2''$	JPL
	$39^{s}.0 \pm 0^{s}.5$	$12'' \pm 5''$	Molonglo
PSR 0950 + 08	$09^{h},50^{m},30^{s}.85 \pm 0^{s}.00$	$08° \; 09' \; 52'' \pm 3''$	JPL
	$30^{s}.65$	$47''.0$	Manch.
PSR 1133 + 16	$11^{h},33^{m},27^{s}.43 \pm 0^{s}.03$	$16° \; 07' \; 31''.4 \pm 0''.9$	JPL
	$27^{s}.1$	$35''.0$	Manch.
PSR 1933 + 16	$19^{h},33^{m},31^{s}.80 \pm 0^{s}.08$	$16° \; 09' \; 58''.5 \pm 1''.2$	JPL
	$10^{s} \pm 20^{s}$	$06' \pm 7'$	Manch.
PSR 2016	$20^{h},16^{m},00^{s}.07 \pm 0^{s}.03$	$28° \; 30' \; 31''.0 \pm 0''.6$	JPL
	$15^{m},58^{s} \pm 20^{s}$		Meudon
		$31' \pm 5'$	AIO
PSR 2045 - 16	$20^{h},45^{m},47^{s}.6 \pm 1^{s}.5$	$-16° \; 28' \pm 2'$	JPL
	$20^{h},45^{m},47^{s}.6 \pm 0^{s}.4$	$-16° \; 27' \; 50'' \pm 12''$	Molonglo

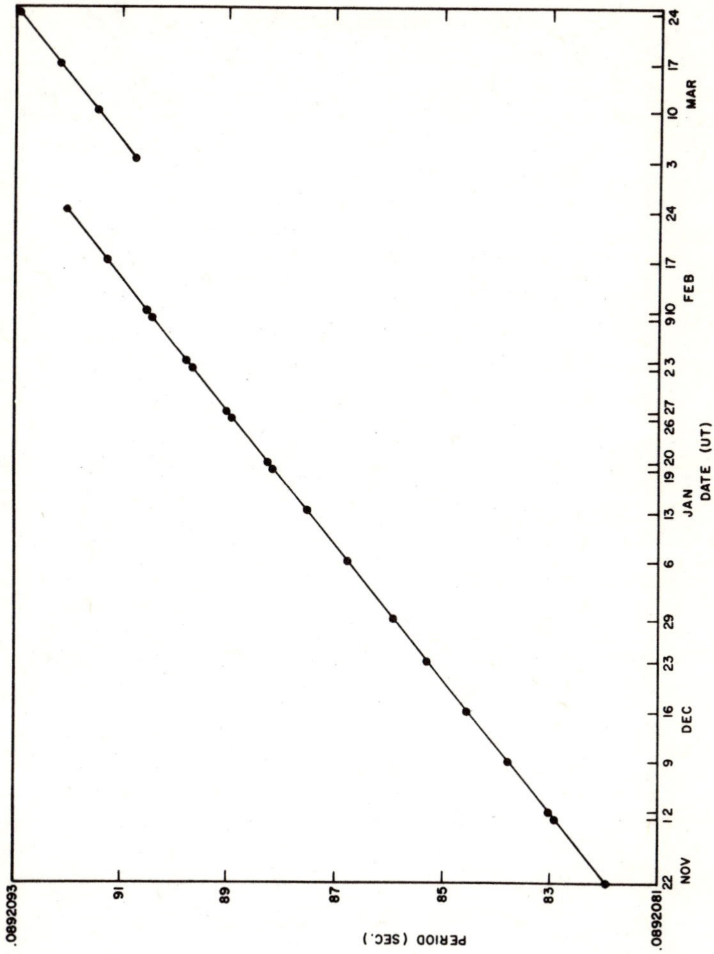

Fig. 4. Barycenter period of PSR 0833-45 as observed from November 22, 1968 to March 24, 1969. The difference between the extrapolated and measured period between February 24 and March 3 is 208 ns.

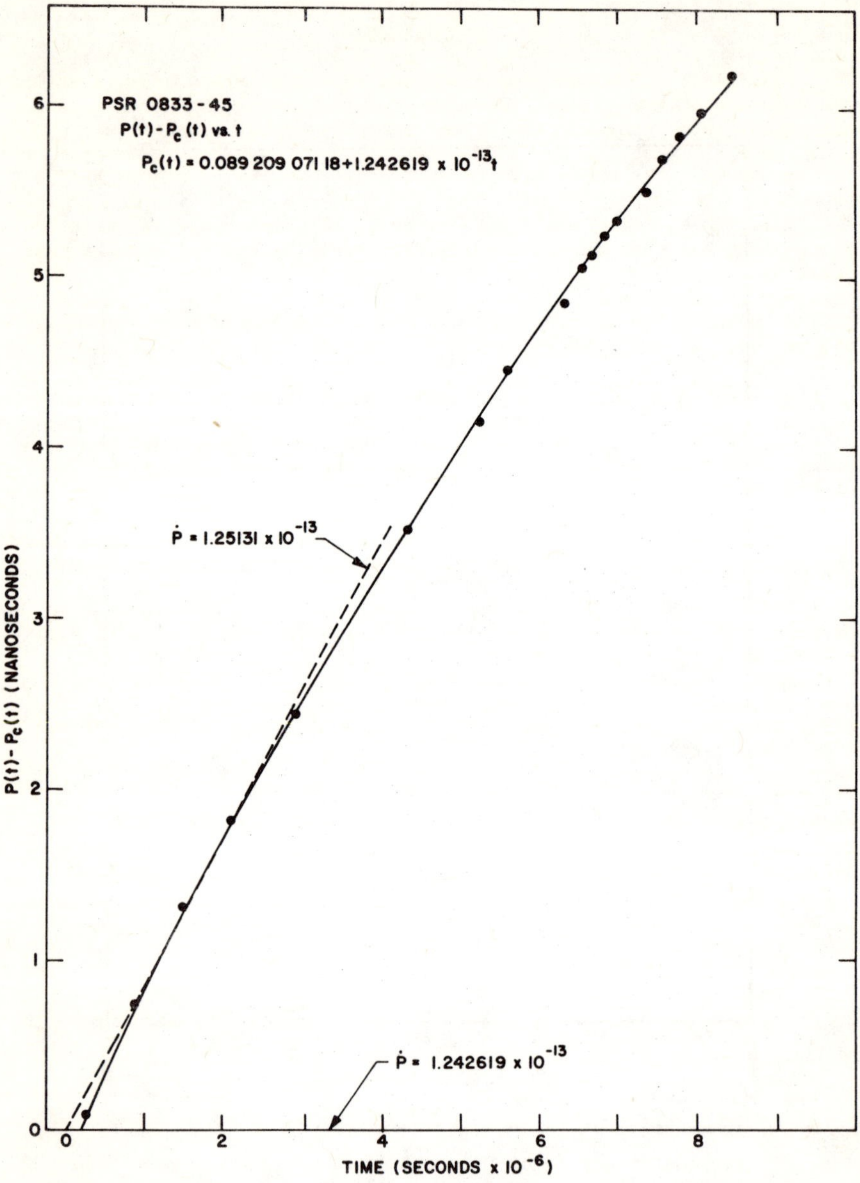

Fig. 5. Reconstructed period of PSR 0833-45. The ordinate is the difference between the actual period and the period predicted by the value of $\dot{P}$ obtained prior to March 3, 1969. The dashed line is the value predicted by the value of $\dot{P}$ obtained just after the period jump.

DISCUSSION

J.A. EARL: "How can you be so sure that the period changes are in the rotation of the pulsar? How can you distinguish that from a period change due to a change in the linear velocity of the pulsar?"

G. DOWNS: "It would have to be a change in the acceleration."

J.A. EARL: "It doesn't have to be an acceleration. It could be going along and just change its velocity maybe because of a comet that comes by.

G. DOWNS: "Yes, I see what you mean. That would make a shift in the period. We can't be sure; there's no way of separating that out."

B. ZUCKERMAN: "How well can you measure position?"

G. DOWNS: "It depends on how well you can measure pulse arrival times. Our measurements are good enough that we are pushing the accuracy of the ephemeris (about 0.1)."

REFERENCES

Boynton, P.E., Groth, E.J., III, Partridge, R.B., and Wilkinson, D.T. 1969, Ap. J. Letters, 157, 197.

Clemence, G.M., and Szebehely, V. 1967, Astron. J., 72, 1324.

Downs, G.S., Morris, G.A., and Reichley, P.E. 1969, Nature, 222, 1257.

Hamilton, W.C. 1964, Statistics in Physical Science, The Ronald Press Co., New York.

Melbourne, W.G., Mulholland, J.D., Sjogren, W.L., and Sturms, F.M. 1968, JPL Tech. Rep. No. 32-1306.

Radhakrishnan, V., and Manchester, R.N. 1969, Nature, 222, 228.

Reichley, P.E., and Downs, G.S. 1969, Nature, 222, 229.

Richards, D.W., Rankin, J.H., and Counselman, C.C., III. 1969, IAU Circ. No. 2164.

October 13, 1969

SEARCHING FOR PULSARS -

RESULTS AND TECHNIQUES

David H. Staelin

Massachusetts Institute of Technology
Boston, Mass.

I shall discuss two topics, techniques for searching for pulsars and a new pulsar model.

Searching for pulsars is basically a problem in strategy, for no search can be general. One guesses the emission properties and spatial distribution of pulsars a priori, and this inevitably biases the statistics of the pulsars that are discovered. I shall review various search techniques and try to estimate the biases which are introduced.

Like many radio sources of interest, pulsars were discovered by accident. Hewish and his co-workers at Cambridge were looking for scintillation in small radio sources when Jocyln Bell, a graduate student, noticed the great regularity of certain pulsating sources of interference.

The major radio searches for pulsars are listed in Table 1. The column headed "Opportunity to be First" indicates whether a group had first access to a significant portion of the sky, either by virtue of their geographic location or by virtue of a new search technique. The great success of the Molonglo Observatory in finding pulsars is due largely to the fact that they made repeated surveys of a large part of the sky, including the galactic plane, and the fact that at 400 MHz they were able to detect weak sources with high dispersion measures. I understand they have approximately 400 suspects of which they believe a fair fraction may be genuine. One reason many suspects are difficult to confirm is that many pulsars turn off for extended periods of time, greater even then six months.

Another factor in Molonglo's favor is that not only is the signal-to-noise-ratio typically better in Molonglo's records because of their large antenna, but also they have two antenna beams slightly off-set in the sky. The displacement amounts to approximately 15 seconds, so that one sees a genuine pulsar twice, first in one beam then in the other. This allows them to discriminate against interference. The source is in one of their beams for only 15 seconds, but only a few pulses in that interval are sufficient to warrant subsequent confirmation.

The simplest computer techniques are based on Fourier transforms of the data. The Fourier transform detects periodicities in the record. Some Fast-Fourier-Transform techniques, such as that of Huguenin and Taylor, obtain additional sensitivity by adding all relevant harmonics for each period being tested.

TABLE 1

STRATEGIES AND TECHNIQUES USED BY DISCOVERS OF PULSARS

Observers	Pulsars Discovered	Opportunity to be First	Strategy	Strength Detectable	Technique
Molonglo	25	Yes	Repeated surveys of large part of sky	Weak	Strip charts, visual examination
Cambridge	6	Yes	Repeated surveys of large part of sky	Strong	Strip charts, visual examination
Arecibo	4	No	Selected suspects Restricted survey	Weak	FFT, FFA
NRAO					
Huguenin-Taylor	3	Yes	Repeated surveys of large part of sky	Strong-Weak	FFT
Reifenstein-Staelin	2	No	Selected suspects One survey, $\sim\pi$ steradians	Weak	FFA, 2-D Chart
Jodrell Bank	1	No	Selected suspects Restricted survey	Weak	Variation of FFA
Russia	1	No	?	?	Strip Chart ?

Notes:
FFT = Fast Fourier Transform
FFA = Fast Folding Algorithm

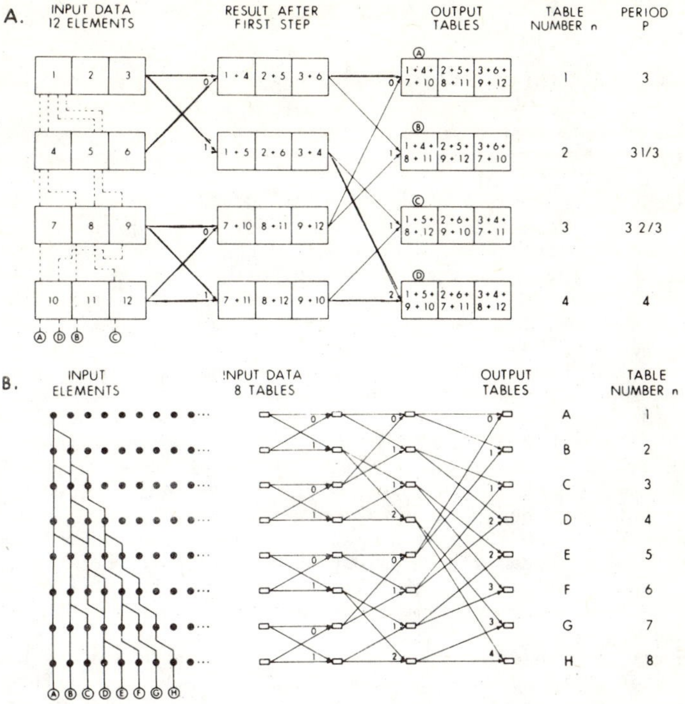

Fig. 1. (a) Flow diagram for fast-folding algorithm as applied to 12 elements. The data are folded at 4 different periods ranging from 3 to 4. The lettered dashed lines clarify the manner in which the elements are combined to form the output tables. (b) A simple generalization of (a) to the case of 8 input tables and arbitrary basic period.

A more sensitive procedure incorporates the Fast-Folding-Algorithm, or FFA, in which a large amount of data is folded on itself in intervals of length equal to the suspected period (Staelin, 1969). All the data at a given phase in the suspected period is thus summed. If there is a pulse with that period it will appear in the summed data. An important feature of this algorithm is its speed. The algorithm functions by adding the data samples in various combinations depending upon the period chosen, and the speed arises because redundant additions are eliminated, as shown in Figure 1. Related algorithms have been used at NRAO, Aricebo, and Jodrell Bank.

In the searches at NRAO made by E.C. Reifenstein and myself, we used the 300-ft transit telescope and a 50-channel radiometer with 100 kHz resolution. On the first pass, we would add the data from 4 or 8 adjacent channels, each channel displaced in time according to the predicted dispersion. The fast-folding principle was used in this analysis too, to avoid redundant additions. The next step was use of the FFA to detect periodicities. Thus our computer simulated a matched filter for dispersed periodic pulse trains. The computer would print out the magnitude of the correlation function, the period at which that correlation function peaked, the sidereal time at which we computed that correlation coefficient, and the dispersion index used for that correlation. We found that with this technique we could process four-minute intervals of data, searching them for two or three thousand periods,

15 phases, and 5 values of dispersion, thus yielding 200,000 points in phase space, and it would take the computer about two minutes to do it. The ratio of telescope time to computer time was 2 to 1, so the computer was busy 5 or 6 hours a day. With this technique we were able to detect almost all the known pulsars at 112 MHz.

In one part of the sky the algorithm yielded an unusual response. Pulsar suspects were indicated at several different periods, but all at the same dispersion. We dumped the 50 channels of data and found that we were indeed observing dispersed pusles. However when we examined the arrival times of these pulses there was no period evident. The reason is now obvious, the pulsar period was about 30 ms and our sample time was 60 ms. The pulsar was observable by us only because of its highly sporadic character. When we plotted the dispersion of these single pulses, the dispersion peaked at two slightly different values. Position measurements indicated two objects 1.2° apart, which we designated NP 0532 and NP 0527. Subsequent observations have shown NP 0532 to be in the Crab Nebula and to have a period of approximately 33 ms. The other one, NP 0527, has a period of 3.74 seconds. I feel that NP 0527 may very well have been ejected from the Crab Nebula explosion, in which case it would be traveling at a speed of about 0.15 c. Fortunately we don't have to speculate very long about this. It should be possible to determine if it's moving that fast by means of an interferometer or accurate pulse timing experiments. The implications of such velocities make this an important experiment.

We also searched for sporadic pulsars. We recorded 50-channel spectrometer data and also displayed it in two dimensions, frequency vs. time, on a memory scope. We would display about two seconds of data and look at it in real time. If we saw even a single dispersed pulse, then we would go back and try to amke a positive detection. This technique is designated the 2-D Display in Table 1. Observations of 59 regions using this technique at 112, 234, and 405 MHz yielded no new pulsars, although the source list included ∿18 supernovae. Searches in these same objects for pulsars with periods as short as 0.010 s were also made unsuccessfully.

Several questions can be asked concerning the results of a pulsar search. For example, are the statistics biased with respect to period, dispersion, or spatial distribution? Is there bias with respect to supernova remnants? Finally, does the search technique introduce bias against certain kinds of sporadic pulsars?

First consider bias with respect to period, dispersion, and spatial distribution. In our search at N.R.A.O. we covered about half the northern sky. In our survey we had trouble detecting pulsars with periods much less than ∿0.2 seconds, or much greater than ∿7 seconds. Our survey was limited primarily to dispersion measures less than ∿150 pc cm^{-3}. We made other surveys which went deeper in dispersion and which encompassed periods as short as 10 ms. However, because those searches involved much more computer time they were restricted to selected objects, primarily supernova remnants.

We find that if you plot pulsar period versus dispersion, as shown in Figure 2, they seem to divide into two groups. Pulsars with periods greater than one second seem to have relatively low dispersion measures while those with periods less than one second have a much wider range of dispersion measures. This effect introduces an additional handicap to detection of short period pulsars.

The Molonglo survey also had difficulty reaching short periods at high dispersion. Their limits were at periods of ∿100 - 200 ms and dispersions of ∿200 pc cm^{-3}. Although their published sensitivity curves don't show any bias against long period pulsars, there is one. This

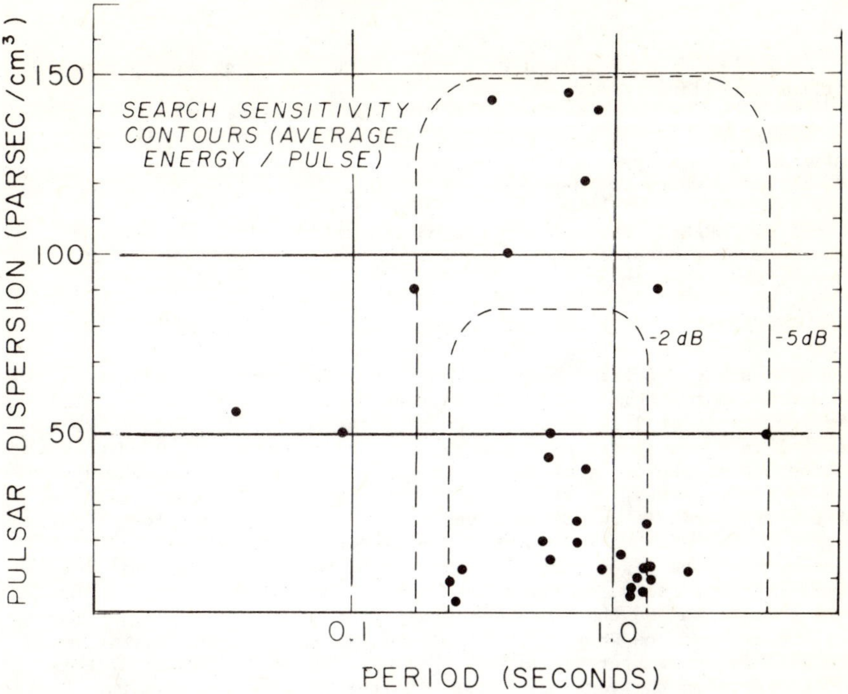

Fig. 2. Distribution of pulsar dispersions and periods. Sensitivity contours for the NRAO 112-MHz search are also shown.

bias is introduced by their beam width which allows them to observe an object for only 15 seconds at 400 MHz. This means it's very hard for them to detect pulsars with periods longer than ∿4 seconds.

Another question to investigate is the association with supernova remnants. Several groups have made searches of supernova remnants including the Molonglo group, the Arecibo group, our own group, and others. Our search of supernova remnants encompassed periods down to 10 ms and, by going to 400 MHz, dispersion measures up to ∿1200 pc cm^{-3}. Molonglo and Arecibo also made special arrangements to look for high dispersion. We also looked for single pulses using a 2-D display, but were unsuccessful. Thus it appears that supernova remnants typically do not contain pulsars. There are only two exceptions, Vela X and the Crab. Therefore, only 5 to 10% of supernova remnants have pulsars in them. Therefore, one must conclude that either most supernova remnants are not associated with the formation of neutron stars or that most neutron stars are not observable pulsars. We also searched Seyfert galaxies, Globular clusters, and some anomalous scintillating low frequency sources with negative results.

Question from B. ZUCKERMAN: "Do you know whether Molonglo searched the Magellanic Clouds?"

STAELIN: "I don't know whether they have or not."

S.P. MARAN: "The declination of the clouds is very close to the limits of what they surveyed but I think they are included."

T.A. MATTHEWS: "Do you know what the dispersion measure would be for the Magellanic Clouds?"

STAELIN: "Well, if you could find one there it would be very interesting. One pulsar has been seen with a dispersion measure of 270 pc cm^{-3}, which corresponds to 10 kpc.

S.P. MARAN: "People have looked optically at the site of the supernova that occured some years ago in the Andromeda nebula but they haven't seen anything."

STAELIN: "That's not too surprising."

Our survey covered about one half the northern sky, and other surveys have covered the same region, as well as most of the rest. Most pulsars are located in or just south of the galactic plane, although there are several in the northern hemisphere. The small number of pulsars in the southern hemisphere may be due to the fact that the southern hemisphere simply isn't covered as thoroughly as the north. If one plots the number of pulsars per unit solid angle versus galactic latitude, and divides them into two groups, P less than 0.8 seconds and P greater than 0.8 seconds than one finds that the long period pulsars are distributed rather isotropically, as shown in Figure 3. The short period pulsars show a distribution that peaks strongly in the galactic plane. This is telling us that the short period pulsars, which are typically younger, are visible at greater distances. They are hotter, brighter objects, but close to us they are more rare than the long period pulsars. The long period pulsars are typically also in the plane, but they are much less bright so one observes only those that are close, and hence they appear isotropic. Another interpretation is that old pulsars are more nearly Population II objects, and the young ones are Population I, but that interpretation is much harder to sustain. No one has fully analyzed all the implications of the distribution of pulsar properties.

How are the searches biased by statistics of pulsar properties? One effect arises because the short period pulsars are discriminated against twice. They are discriminated against because of their short period and also because they are highly dispersed. Therefore the short period pulsars are probably much more highly concentrated in the galactic plane than Figure 3 indicates. On the other hand the long period pulsars are slightly discriminated against because if they are sporadic, they can easily be missed.

Question from C.E. FICHTEL: "How far away are the long period pulsars?"

STAELIN: "The dispersion measures indicate that they are on the average 400 pc away, in contrast to the short-period pulsars which average $\sim$2 kpc."

Before discussing our model for the radio emission from a pulsar, let me remind you of the three characteristic features of pulsars: 1) The radiation is partially polarized with the plane of polarization rotating through the pulse, 2) there are subpulses which may march through the pulse, generally from back to front, and 3) average pulse shapes are generally double with the separation between pulses becoming wider at lower frequency. The $f^{-1/4}$dependence of the width that is sometimes quoted is not found in all pulsars -- in some the dependence

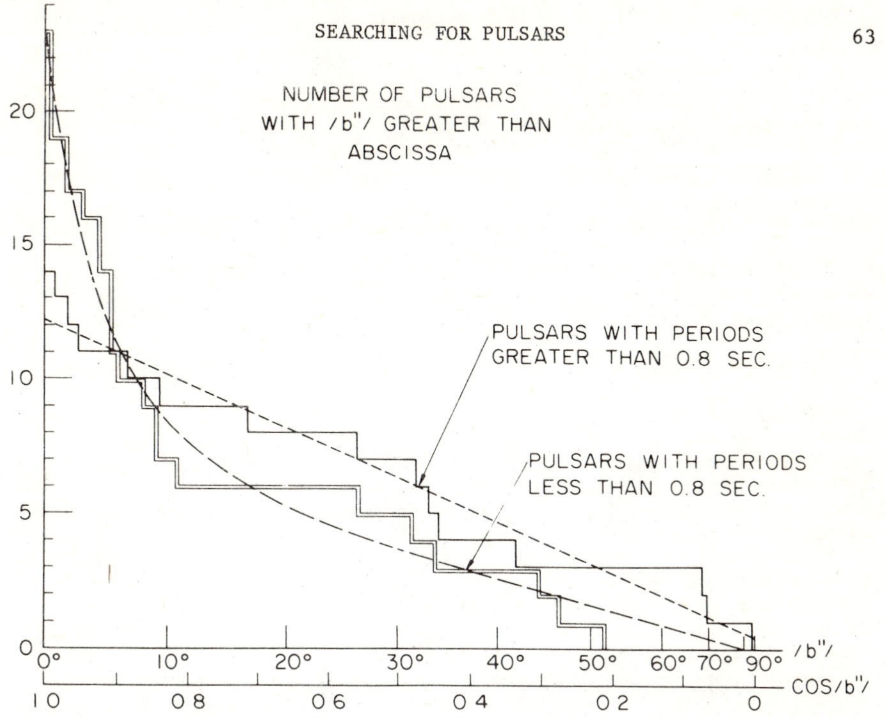

Fig. 3. Distribution of pulsars in galactic latitude showing tendency for short period pulsars to concentrate near the plane, and long period pulsars to be isotropic.

is stronger than this. Any emission mechanism must be consistent with these properties as well as allowing for the variation from one pulsar to another.

One thing that everyone agrees upon is that at the core of a pulsar there is probably a spinning neutron star. The neutron star probably has a magnetic field, say a dipole field in the first approximation. Surrounding the neutron star there will be a very hot plasma. Gunn and Ostriker have argued, very convincingly I think, that as the star spins the low frequency magnetic dipole radiation is very substantial. There is thus a large amount of energy propagated outward at low frequency. They argue that as these waves propagate outward they become TEM waves in the far field and that these waves are very efficient in accelerating charged particles. If the crossed electric and magnetic fields of the wave are sufficiently intense they will make a charged particle ride outward on the crest of the wave much like a surfboard riding a wave. The particle will then travel at the velocity of the wave. In their theory this produces electrons with energies with 10^{13} - 10^{15} eV.

What is the ultimate destination of this tremendous flux of outwardly moving relativistic particles? It's reasonable to assume that the hot, sparse plasma near the star is surrounded by a relatively cool, dense plasma. There will be some boundary, more or less sharp, at which pressure equilibrium is maintained. The particles will therefore penetrate the denser plasma, and many mechanisms have been hypothesized for what might happen there.

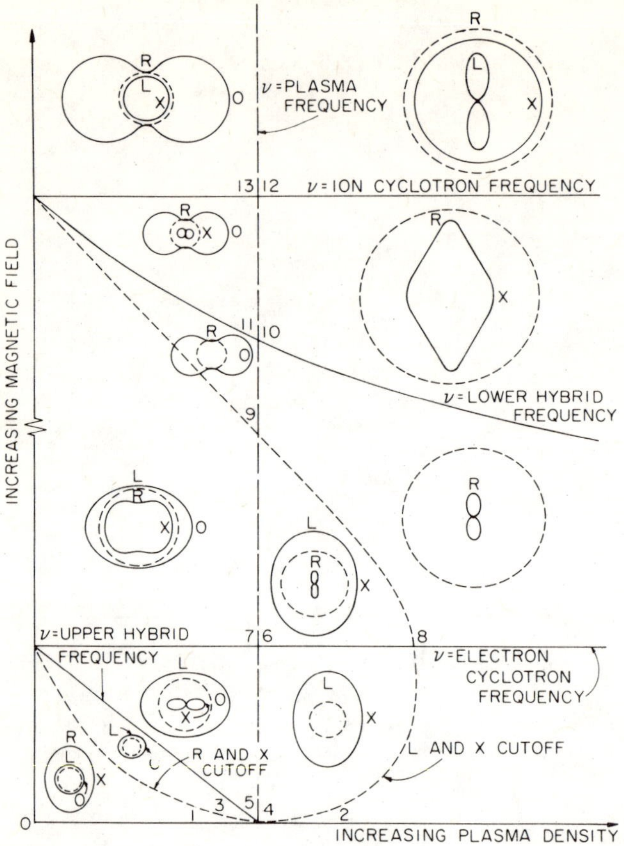

Fig. 4. CMA diagram showing phase-velocity surfaces for cold-plasma electromagnetic waves at frequency ν, as a function of magnetic field strength and plasma density. For a given mode phase velocity is proportional to the diagram radius vector, where the dotted circles correspond to the velocity of light. The magnetic field is parallel to the abscissa. R, L, X and O represent right-circular, left-circular, extraordinary, and ordinary waves, respectively.

We began by asking what kind of waves can propagate in a given magnetized plasma. The CMA diagram is particularly useful when considering a cold plasma. Such a diagram, in Figure 4, shows parameter space divided by the many characteristic frequencies of a plasma. Many of these modes have phase velocities less than c; in particular, every region with number greater than 4 has one slow mode. Slow waves are most common at frequencies below the electron cyclotron resonance, for which the extraordinary mode propagating across the magnetic field has a phase velocity less than the speed of light. The particles traveling at nearly the speed of light will therefore generate Cherenkov radiation. The radiation generated will be right circularly polarized or in general eliptically polarized. If one considers a pulsar which has an upper frequency cut-off of about 1 GHz, as several do, it follows that the magnetic fields in the plasma layer are greater than about 1000 or 2000 gauss. In order to radiate appreciably at say 10,000 MHz, pulsars require

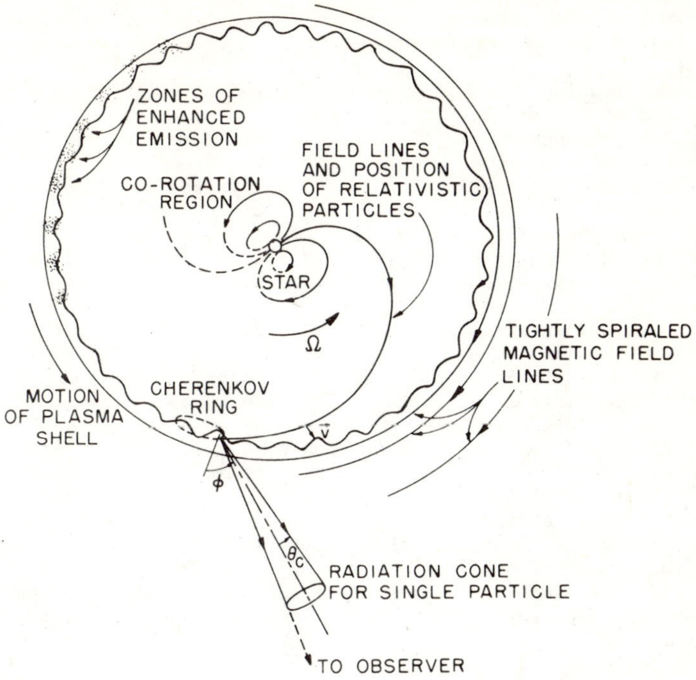

Fig. 5. Example of pulsar emission model involving 1) circum-pulsar plasma waves to produce marching subpulses, and 2) Cherenkov processes to yield other pulsar emission properties. Solid field lines refer to hemisphere appropriate to observer, and dotted lines to other hemisphere.

very high electron cyclotron frequencies, and hence still more intense fields. No pulsars radiate appreciably at such frequencies.

To summarize, the model contains a star with a field which approximates a magnetic dipole as shown in Figure 5. The particles are injected into the outward-moving mode by mechanisms such as those of Goldreich and Julian, and/or Gunn and Ostriker. The particles move radially outward and at any instant of time lie along a spiral. The radial motion will be slightly modified by a tendency of the particles to acquire some of the angular momentum of the star, so that when the particles hit the denser plasma they will be moving slightly forward as well as outward. The Cherenkov radiation will be generated in the usual cone. Summing across the distribution of particles will smear this cone slightly. The semiangle of this cone is given by $\beta n \cos \theta_c = 1$ where $\beta = v/c$, and n is an index of refraction for the mode in question. This cone of radiation then rotates about the star with the same period as the star itself. The radiation will typically be elliptically polarized. The geometry will make the plane of polarization rotate during the course of the pulse. There may also be some Faraday rotation contributing to the rotation of the plane of polarization.

The first feature of pulsar emission we wished to explain concerned the polarization and its systematic variation during the pulse, and it is inherent in this model. The Cherenkov cone accounts for the double pulse shape. The wavelength dependence of the separation between the two peaks will vary from one pulsar to another simply because the dispersion relation which is appropriate will vary. This is a natural consequence of the model.

Another characteristic feature, the subpulses and particularly the marching of the subpulses from the back to the front of the pulse, is contained in this model. Instabilities should occur in the boundary layer. If, for example, the radius of the boundary layer is several times the radius of the light cylinder, and if the radiation cone is canted slightly forward, then any standing wave built up in the plasma containing zones of enhanced emission can move through the Cherenkov ring producing marching subpulses. The Cherenkov ring is the locus of points on the plasma shell from which one would see radiation.

In order for the pulsar radiation to be received, two conditions are required. First the Cherenkov radiation must be generated, and second, the radiation must convert to a mode which can escape the star. Mode coupling is necessary because slow waves cannot cross all the boundaries of phase space and excape to free space. They must couple to modes that can cross boundaries. However this is not hard to do. The radiation can scatter from one mode into the orthogonal mode. This scattering may take place at inhomogenities. Even in a homogeneous plasma there can be coupling from one mode to another because of non-linear effects. This coupling process will depend upon the parameters of the plasma, and any waves in the plasma will modulate this coupling. These zones of enhanced emission presumably rotate in the same direction as the star, but slightly slower than the magnetosphere. Thus the effect of a particular enhancement is first evident at the trailing edge of a pulse, and it later passes through the Cherenkov ring and disappears at the leading edge of the pulse window.

The original interpretation of the subpulses was that they reflected a second periodicity modulated by the pulsar. Subsequent observation has shown that this is not correct, and that we are simply observing subpulse features moving across the pulse at a slightly variable rate. This motion is not stable and is not a high-Q process.

Question from S.P. MARAN: "What's the argument for it not being a modulation of one period by another?"

STAELIN: "The subpulses always march in the same direction at about the same rate. If you had a second period that was being modulated you would expect it to march through at an arbitrary angle, and that's not what happens."

In Cherenkov-like processes the radiation from an individual particle is coherent, whereas the individual particles may be coherent or incoherent. Many people have argued that because of the very high surface brightness of pulsars, the radiation cannot be incoherent. These surface brightnesses are calculated by taking the briefest observed pulse, say 0.1 ms, and assuming that the emitting region is no bigger than the distance light can travel in that time. This gives dimensions of 30 km and leads to brightness temperatures that imply unreasonable particle energies of the order of 10^{25} eV. Hence it must be a coherent process. This argument does not apply to the Cherenkov model, because the emitting region can be substantially larger than

30 km. Consider a "wall" of particles moving radially outward. When they hit the plasma shell each particle or clump of particles creates its own Cherenkov cone of radiation. Because the particles and the photons they produce both move in approximately the same direction at velocity c, the duration of a subpulse, say 0.1 msec, is independent of the horizontal extent of the wall. It is determined by the thickness of the wall along the line of sight. For example, assume that the plasma shell in CP 1919 is at 10 times the light-cylinder radius, i.e., 10^9 meters. Then electrons with energies 10^{13} to 10^{15} eV can supply the surface brightnesses required. These are just the particle energies calculated by Gunn and Ostriker.

Question from H. OGELMAN: "In NP 0532 both the optical and the X-rays originate at the same time and I can't quite see how that's possible in your model."

STAELIN: "Recent unpublished data from Arecibo shows that the typical pulse shape in the radio region from NP 0532 has a spike about 150 μs wide preceded ∿2 msec earlier by a pulse about 2 ms wide. The interpulse is again a strong narrow peak resembling the main pulse. In the optical and in the X-ray there are only the two sharp peaks. The broad bump in front of the main pulse is missing. The spikes are aligned within 150 μs and they may be aligned much better than that. The spikes are also fairly stable. The broad feature that precedes the main spike is more variable. The Cherenkov mechanism which I've been discussing may correspond to the broad precursor feature, and the sharp spikes may correspond to another radiation mechanism. That is, the spikes which are well aligned may originate in the same region of space separate from that of the precursor and may be produced by another mechanism. Note that the "wall of particles" picture is consistent with coincident pulses arising from points radially displaced substantial distances, simply because the velocities of the photons and the particles are so similar."

H. OGELMAN: "Can you give a spectrum?"

STAELIN: "It's difficult because many things affect it. The most certain aspect is the upper limiting frequencies, because the pulsar can't emit at frequencies that are so high that the plasma wave propagates faster than the particles."

REFERENCES

Staelin, D. 1969, Proc. IEEE, 57, 724.

October 27, 1969

INTERSTELLAR CLOUD PROPERTIES REVEALED BY PULSARS

Robert M. Hjellming

National Radio Astronomy Observatory*
Green Bank, West Virginia

Most of the time astronomers can only investigate and interpret the properties of radiation emitted by sources in the sky. In some cases the radiation from an emitting source is affected by passage through the intervening medium and we can then obtain information about that medium. This is very important for studies of the interstellar medium because interstellar gases are rather parsimonious in the emission of radiation. If, a few years ago, astronomers had been asked what type of radiation sources would be best for probing the interstellar medium, the wise ones would have asked for large numbers of point sources emitting large amounts of linearly polarized radiation in periodic pulses of short duration. The pulsars have these properties and give us a chance to make some very straightforward measurements. My purpose today is to discuss what can be learned about the interstellar medium from various measurements of pulsar radiation.

1. Dispersion Measure

The earliest pulsar observations by Hewish et al., (1968) established that the arrival time of individual pulses varies with the frequency of observation. These time delays behave exactly as expected for dispersion by electrons in a plasma; they are functions only of the frequency of observation and the integral of the electron concentration along the line of sight, $\int n_e ds$. This integral, the dispersion measure, is the primary observational parameter obtained from measurements of pulsar radiation. If we define

$$\langle n_e \rangle = \frac{\int_0^d n_e \, ds}{d} \tag{1}$$

for a pulsar at a distance d, then a knowledge of d allows a determination

*Operated by Associated Universities, Inc., under contract with the National Science Foundation.

of $\langle n_e \rangle$, and vice versa. The discovery of the pulsar NP 0532 in the Crab Nebula by Staelin and Reifenstein (1968) provided the first good chance to use this relationship. Knowing that d = 2020 pc (Trimble, 1968) for the Crab Nebula and $\int n_e ds = 56.88$ pc cm^{-3} for the Crab pulsar, $\langle n_e \rangle = 0.028$ cm^{-3}. It has been suggested (Large et al., 1968)that the pulsar PSR 0833 is associated with the Vela supernova remnant for which a probable distance of 500 pc is given by Milne (1968). Since $\int n_e ds = 67$ pc cm^{-3} (Manchester et al., 1969) a simple application of equation (1) would give $\langle n_e \rangle = 0.14$ cm^{-3}; however, Manchester et al. (1969) have shown that the Gum nebula may contribute 80 percent of the observed dispersion so that $\langle n_e \rangle = 0.03$ cm^{-3} for the remainder of the line of sight. Unfortunately the correction for the effects of the Gum nebula is very uncertain so any results based on measurements of PSR 0833 will be very uncertain.

2. Hydrogen Absorption

A second important observational parameter is obtained from measurements of the 21-cm absorption of the pulsar signals. The nature of the pulsar radiation makes this measurement very elegant. A schematic illustration of the measurement process is shown in Fig. 1. First you observe the line when the pulsar is turned off to determine the emission profile for all the gas in the beam of the antenna. Then, when the pulsar is turned on, a nearly uniform increase of intensity at all frequencies in the line would be observed if there were no absorption of the pulsar signal. In some cases the increase is not uniform and the difference between a uniform increase and the observed increase gives the absorption profile for all frequencies. The observations determine τ_ν for all frequencies and hence the integral $\int \tau_\nu d\nu$. If n_H = neutral hydrogen concentration and T = temperature, then

$$\int_o^d (n_H/T)\, ds = 3.85 \times 10^{14} \int_{\text{line}} \tau_\nu d\nu \qquad (2)$$

where c.g.s. units are used. The integral $\int (n_H/T)ds$ is the second important observational parameter obtained from pulsar observations; this integral contains information about both the neutral hydrogen concentration and the temperature.

Now, if the interstellar medium were uniform, the ratio $\int n_e ds / \int (n_H/T)ds$ would be nearly the same for all lines of sight. This is definitely not true as may be seen from Table 1 where $\int n_e ds$, $\int \tau_\nu d\nu$, and $\int n_e ds / \int (n_H/T)ds$ are presented for the four pulsars for which all of this data is known. The value of $\int \tau_\nu d\nu$ for NP 0532 is taken to be that measured for the Crab Nebula by Clark (1965). The data for CP 0328 and PSR 1749 is taken from Gordon et al. (1969) and the data for MP 0736 is taken from Manchester et al. (1969). The interstellar gas is clearly not uniform, but we cannot yet conclude whether variations of n_e, n_H, T, or any combination of these three, are responsible for the variation found in Table 1.

Unfortunately, attempts to measure 21-cm absorption in the signals from other pulsars have mainly produced only upper limits to the optical depth. The results listed in Table 2 show the galactic longitude (ℓ^{II}), galactic latitude (b^{II}), $\int n_e ds$, maximum optical depth (τ_m), and the ratio

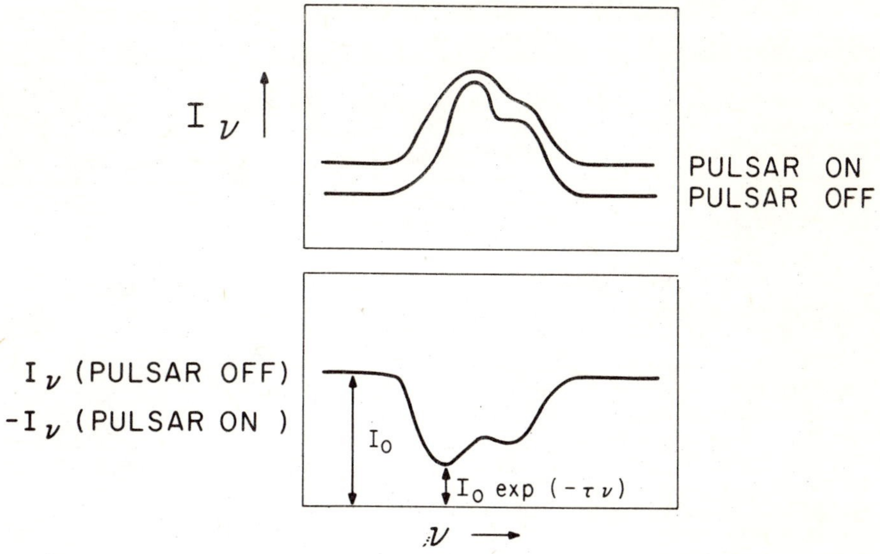

Fig. 1. Schematic illustration of 21-cm absorption line measurement with pulsar signals. The upper diagram shows the emission profile for the gas as observed between pulses ("pulsar off") and during a pulse ("pulsar on"). The lower diagram shows the difference between the two profiles, i.e., the absorption profile.

TABLE 1

PULSAR ABSORPTION DATA

Pulsar	$\int n_e ds$ [pc cm^{-3}]	$\int \tau_\nu d\nu$ [kHz]	$\int n_e ds / \int (n_H/T) ds$ [°K]
NP 0532	56.88	42	10.9
CP 0328	26.75	74	2.9
PSR 1749	50.88	18.9	21.6
MP 0736	100.0	71	11.3

TABLE 2

RESULTS OF PULSAR ABSORPTION SEARCHES

Pulsar	ℓ^{II}	b^{II}	$\int n_e ds$ [pc cm^{-3}]	τ_m	$\tau_m/\int n_e ds$ [pc^{-1}cm^3]
PSR 1749	1.5	-1.0	50.9	0.75	0.015
CP 0328	145.0	-1.2	26.8	1.2 0.75	0.045 0.028
MP 0736	253.7	-8.9	100.0	>>1	>>0.01
PSR 1929	47.4	-3.9	8	<0.1	<0.012
JP 1933	52.3	-2.0	143	<0.5	<0.0035
NP 0527	183.8	-6.9	49.3	<0.5	<0.01
AP 0823	197.0	+31.8	19.4	<0.4	<0.02
CP 0950	228.9	+43.7	2.98	<0.3	<0.1
CP 1133	241.9	+69.2	4.87	<0.3	<0.06
PSR 0833	263.5	-2.8	69	<0.1	<0.0014

$\tau_m/\int n_e ds$ for the pulsars listed in Table 1 and seven other pulsars* observed by Gordon and Gordon (1969). Because τ_ν is proportional to n_H/T, the ratio $\tau_m/\int n_e ds$ should be a roughly distance-independent estimate of the ratio of the amount of cool hydrogen to the amount of electrons along the line of sight. We see that there are at least two lines of sight, towards JP 1933 and PSR 0833, which are unusually lacking in cool hydrogen. Further kinematical information concerning the radial velocities of absorbing clouds can be obtained from the pulsar absorption experiments; but we will not discuss the application of these results.

3. Average Magnetic Fields

The third parameter of the interstellar gas that may be obtained from pulsar observations is the integral $\int n_e B_\parallel ds$, where $B_\parallel$ is the component of the magnetic field vector along the line of sight. This can be obtained for the pulsars which have measurable linear polarization

*The maximum optical depths are given in terms of 2σ limits.

because the position angle of linearly polarized radiation changes by an angle $\Delta\theta$ when it passes through a plasma, where

$$\Delta\theta \quad \alpha \quad \lambda^2 \int n_e B_{\parallel} \, ds \ . \tag{3}$$

In equation 2, λ is the wavelength at which the measurement is made; when $\Delta\theta$ is known for several wavelengths the value of $\int n_e B_{\parallel} \, ds$ is determined. An average longtitudinal magentic field is then found from $\langle B_{\parallel} \rangle = \int n_e B_{\parallel} ds / \int n_e ds$. The results obtained for seven pulsars summarized by Verschuur (1969) are shown in Table 3. In Table 3 a positive sign for $\langle B_{\parallel} \rangle$ indicates a direction away from the observer. We see that there are substantial variations in $\langle B_{\parallel} \rangle$, but without more information the variation may be due to any combination of variations of field directions and field strength. However, these fields are rather small compared to the values of 20, and even 50 μG, found by Verschuur (1969) in certain cool clouds.

4. Interstellar Scintillation

Observations of pulsars show that there are slow variations in the amplitude of the pulses which have an irregular structure in frequency. The narrower the frequency bandwidth of the observation is made, the greater the fluctuations become. Rickett (1969) has defined a characteristic bandwidth, $\Delta\nu$, as that receiver bandwidth which yields r.m.s. fluctuations that are a factor of two smaller than those corresponding to $\Delta\nu = 0$. Richett (1969) has shown that the $\Delta\nu$ is proportional to $(\int n_e ds)^{-2}$ as predicted from scintillation theory; therefore the variations in pulse amplitude must be produced by the same electrons which produce the dispersion. The mere fact of scintillation then implies that dense clumps of electrons must exist along the line of sight. We will not discuss any further details of scintillation theory of interpretation.

TABLE 3

AVERAGE FIELDS IN THE DIRECTION OF PULSARS

Pulsar	ℓ^{II}	b^{II}	$\int n_e ds$ [pc cm^{-3}]	$\langle B_{\parallel} \rangle$ [10^{-6} Gauss]
AP 2015	68.1	-4.0	14.2	+2.0
CP 0328	145.0	-1.2	26.8	+3.5
NP 0527	183.8	-6.9	49.3	|0.9|
NP 0532	184.5	-5.8	56.9	+0.6
PSR 0833	263.5	-2.8	63	-0.8
CP 0808	139.9	31.6	5.8	+2.3
CP 0950	228.9	43.7	2.98	<0.2

5. Cosmic Ray Ionization of the Interstellar Gas

As a prelude to making explicit use of the data on $\int n_e ds$ and $\int (n_H/T) ds$ let us discuss the only existing quantitative theory that predicts the relation among T, n_e, and n_H in the interstellar gas. This theory is based upon assuming that low energy cosmic rays ($E \lesssim 50$ MeV/nucleon) are responsible for most of the ionization and heating in the interstellar gas. This process involves direct transfer of kinetic energy of cosmic rays to the bound electrons of hydrogen and helium by direct Coulomb interactions. When this interaction produces ionization the kinetic energy of the electrons is partially converted into thermal energy in the gas. By assuming that the resulting weakly ionized plasma is in a state of statistical equilibrium, we can write down equations saying that (1) the energy input to the gas is exactly balanced by energy loss due to cooling processes and (2) each ionization rate is exactly balanced by the rate of the reciprocal electron capture process. These equations involve only four physical variables: T, n_e, n_H, and the total ionization rate per hydrogen atom, which we denote by ζ. The equations can be solved (Hjellming et al., 1969, Spitzer and Scott, 1969) to uniquely determine the relation between T, n_e/n_H, and ζ/n_H. In addition, if we define p_g to be the gas pressure, assuming the perfect gas law, the ratio p_g/ζ depends only on T, n_e/n_H, and ζ/n_H. Thus n_e/n_H, ζ/n_H, and p_g/ζ can all be regarded as functions only of T; the results of computations of these relationships, which are discussed in detail by Hjellming et al., (1969), are shown in Fig. 2.

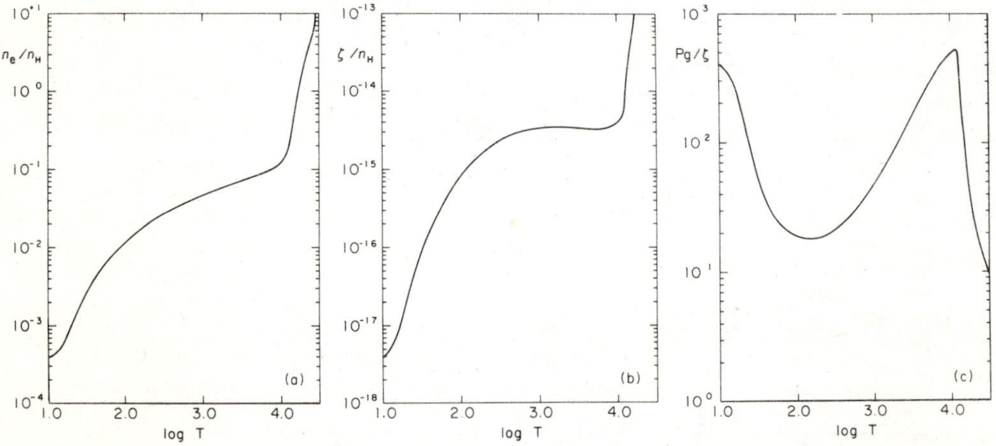

Fig. 2. Equilibrium conditions of interstellar gas for cosmic ray ionization (Hjellming et al., 1969).

6. Interstellar Clouds and a Uniform Inter-Cloud Medium

As discussed earlier, we know that the interstellar medium is very non-uniform. In order to carry out quantitative calculations we then need to adopt a model which makes some allowance for inhomogeneities. Therefore let us assume that each line of sight contains: (1) cool clouds with parameters $(n_e)_c$, $(n_H)_c$, T_c, and a total thickness Δs; and (2) a "uniform" inter-cloud medium with parameters $(n_e)_i$, $(n_H)_i$, and T_i. This model is based upon observations indicating that the major inhomogeneities are cool clouds ($T_c \sim 20$ to 60 °K).

Even this two-component model will be useful only if we can assume some sort of coupling between the clouds and the inter-cloud medium; thus we assume that $(p_g)_c$ for each cloud is related to $(p_g)_i$ according to

$$(P_g)_i = (p_g)_c + \Delta p \tag{4}$$

where Δp may be different for each cloud. If Δp is considered to express the gradients in magnetic pressure, turbulent pressure, and cosmic ray pressure, then equation (4) represents an assumption of pressure equilibrium. However, if we consider Δp to be a free parameter, equation (4) could represent the current state of an evolving cloud.

In order to make practical use of equation (4) we must assume that ζ, the ionization rate per hydrogen atom, is constant along the line of sight. We then can write

$$\frac{(P_g)_i}{\zeta} = \frac{(p_g)_c}{\zeta} + \frac{\Delta p}{\zeta} \tag{5}$$

and we can use Fig. 2c to relate the clouds to the inter-cloud medium; the ratio $\Delta p/\zeta$ is then taken to be a free parameter.

Unfortunately the model as we have desribed it contains two possible choices for the inter-cloud medium. This is shown schematically in the p_g/ζ vs T diagram in Fig. 3. We begin by assuming that the cool gas phase is to be associated with cool clouds, that is, $T_c < 130°$ K. However, for every value of T_c, there are two possible values of T_i that will give the same value of p_g/ζ; we will refer to these two possibilities as the 10^3 °K phase, which is thermally unstable (Field et al., 1969), and the 10^4 °K phase, which is thermally stable. Field et al., (1969) and others have chosen the 10^4 °K medium to avoid the theoretical problems associated with thermal instability. However, some observations exist which indicate that T_i in the galactic plane cannot be as high as 10^4 °K. Some emission line profiles measured at low galactic latitudes have widths indicating that the temperature of the emitting gas must be much less than 10^4 °K. In addition, the directly observed harmonic mean temperature for interstellar hydrogen is in the range 125 - 150 °K; it would be very difficult to obtain this harmonic mean temperature by combining contributions of 20 to 60 °K cool clouds and a 10^4 °K inter-cloud medium, but it would be very easy to obtain this with reasonable numbers if the inter-cloud temperatures are of the order of 10^3 °K. Hence we will assume that $130 < T_i < 10^4$ °K and we will have to worry about the consequences of a thermally unstable inter-cloud medium.

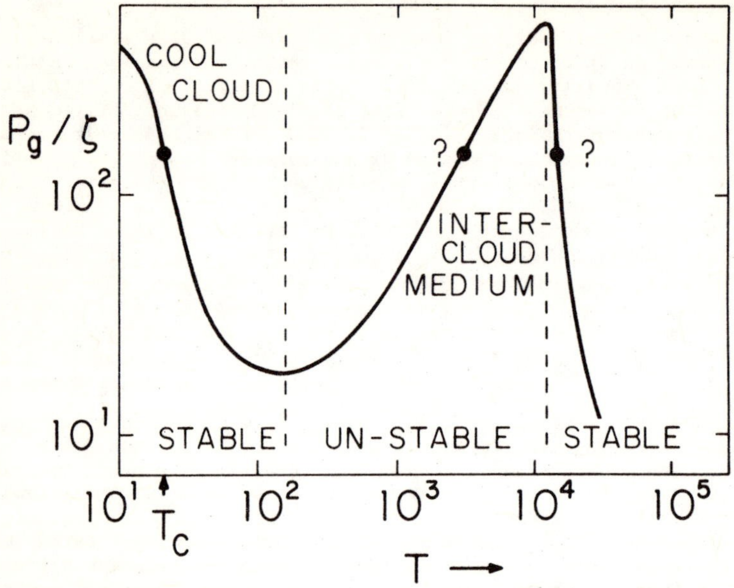

Fig. 3. Pg/ζ vs T diagram, showing that the model described here contains two possible choices for the inter-cloud medium.

Once we choose what type of inter-cloud medium we will discuss, we can then relate in a simple fashion most of the parameters of the two component model. We will take an approach whereby we adopt a value of T_i, then using the results of Fig. 2, $(n_e)_i/(n_H)_i$, $\zeta/(n_H)_i$, and $(p_g)_i/\zeta$ are determined. If we further choose a value of $\Delta p/\zeta$, we can determine $(p_g)_c/\zeta$ from equation (5), and then T_c, $(n_e)_c/(n_H)_c$, and $\zeta/(n_H)_c$ are determined using Fig. 2. Knowing these the ratios $(n_e)_c/(n_e)_i$ and $(n_H)_c/(n_H)_i$ may be calculated. The results for a range of values of T_i are presented in Table 4 for the special case where $\Delta p = 0$; similar tables may be calculated for any value of $\Delta p/\zeta$.

If we write down the expression for $\int n_e ds / \int (n_H/T) ds$ for the two component model, we can solve to obtain

$$\frac{\Delta s}{d - \Delta s} = \frac{(n_H)_i}{(n_H)_c} \frac{\left[\frac{(n_e)_i}{(n_H)_i} - \frac{1}{T_i} \frac{\int n_e ds}{\int (n_H/T) ds} \right]}{\left[\frac{1}{T_c} \frac{\int n_e ds}{\int (n_H/T) ds} - \frac{(n_e)_c}{(n_H)_c} \right]} \qquad (6)$$

so that $\Delta s/d$ can be determined for any values of T_i and $\Delta p/\zeta$. A similar manipulation of $\int n_e ds$ gives

$$(n_e)_i = \frac{\int n_e ds}{d\left[1 + \frac{(n_e)_c}{(n_c)_i}\frac{\Delta s}{d}\right]} \tag{7}$$

Because of equations (6) and (7), a knowledge of $\int n_e ds$, $\int (n_H/T)ds$, and d allows a determination of $(n_e)_i$, and once this is known, then $(n_H)_c$, $(n_H)_i$, $(n_e)_c$, Δs, and ζ are all determined for each value of T_i and $\Delta p/\zeta$.

The line of sight to the pulsar NP 0532 in the Crab Nebula is the only certain case for which we can calculate all the parameters of a two-component medium. The results are presented in Table 5 for values of T_i = 1000, 2000, and 3000 °K and $\Delta p/\zeta = 0, \pm 100$.

The effects of different values of Δp, as found in Table 5, can be discussed in terms of the schematic diagram in Fig. 4. If $\Delta p > 0$, so the extra pressure gradient would tend to help the clouds expand, the resulting cloud temperatures are hotter, but cannot exceed $\sim$ 130° K or the cloud would dissipate. On the other hand, if $\Delta p < 0$ so the extra pressure gradient tends to help compress the cloud, the cloud temperatures are lower than they would be otherwise, but if Δp is too negative the value of T_c would reach a lower limit ($\sim$10 - 20 °K, a very uncertain number) after which cloud compression would occur.

Although all of the assumptions that have been made are necessary to obtain quantitative results, it should be remembered that two of them are rather questionable. There is no evidence that ζ is constant over long distances and it may be very variable if the low energy cosmic ray sources are dispersed in space. The second questionable assumption is that of a uniform inter-cloud medium.

7. Evidence for a Non-Uniform Inter-Cloud Medium

There are several sources of evidence that indicate that the most doubtful assumption that we have made is that of a uniform inter-cloud medium:

(1) The instability in the inter-cloud medium should produce fluctuation in density.

(2) Low frequency radio observations in the galactic plane (Bridle and Venugopal 1969) indicate that with $T_i \sim 10^3$ °K, it is necessary that $\langle n_e^2\rangle/\langle n_e\rangle^2 \sim 10$. This is possible only with substantial variations in n_e.

(3) The mere fact of scintillation of the pulsar signals indicates great variations in the inter-cloud medium. We know that the inter-cloud medium produces nearly all of the dispersion, and the dispersing gas is identical to the gas producing the scintillations, hence the inter-cloud medium must be very "lumpy."

(4) Various studies (Rohlfs et al., 1969) have indicated that at high galactic latitudes $n_e/n_H \gtrsim 0.1$ and $T \gtrsim 10^4$; hence there are substantial variations in T_e, n_e, and n_H with distance from the galactic plane. This has the interesting consequence that above a certain distance from the plane of the galaxy, the gas is thermally stable.

TABLE 4

PROPERTIES OF A TWO COMPONENT INTERSTELLAR MEDIUM WITH $\Delta p = 0$

T_c [°K]	$(n_e)_c/(n_H)_c$	$\zeta/(n_H)_c$ [$cm^3\ sec^{-1}$]	P_g/ζ [$dyne\ cm^{-2}\ sec$]	T_i [°K]	$(n_e)_i/(n_H)_i$	$\zeta/(n_H)_i$ [$cm^3\ sec^{-1}$]	$(n_H)_c/(n_H)_i$	$(n_e)_c/(n_e)_i$
15	0.00057	7.40×10^{-18}	300.0	5900	0.091	3.30×10^{-15}	450.0	2.8
20	.00104	2.08×10^{-17}	148.7	2950	.071	3.35×10^{-15}	161.0	2.4
30	.00233	8.17×10^{-17}	57.4	1200	.052	3.45×10^{-15}	42.2	1.9
40	.00375	1.76×10^{-16}	35.3	710	.042	3.20×10^{-15}	18.2	1.6
50	.00501	2.72×10^{-16}	28.1	530	.036	3.01×10^{-15}	11.1	1.5
60	.0063	3.86×10^{-16}	23.7	400	.032	2.7×10^{-15}	7.0	1.4
70	.0078	4.90×10^{-16}	21.6	340	.028	2.48×10^{-15}	5.1	1.4
80	.0091	6.00×10^{-16}	20.0	280	.025	2.22×10^{-15}	3.7	1.3

TABLE 5

LINE OF SIGHT TO NP 0532

T_c [°K]	d	T_i [°K]	$(n_H)_c$ [cm^{-3}]	$(n_H)_i$ [cm^{-3}]	$(n_e)_c$ [cm^{-3}]	$(n_e)_i$ [cm^{-3}]	Δs [pc]	ζ [s^{-1}]	$\int n_H ds$ [pc cm^{-3}]
					$\Delta p/\zeta = 0$				
33.8	2020	1000	18.4	0.591	0.0513	0.0281	7.56	1.99×10^{-15}	1330
24.2	2020	2000	41.0	0.455	0.0613	0.0281	2.87	1.57×10^{-15}	1030
20.0	2020	3000	67.0	0.398	0.0687	0.0281	1.51	1.36×10^{-15}	904
					$\Delta p/\zeta = 100$				
32.3	2020	3000	14.2	0.397	0.0367	0.0281	11.5	1.36×10^{-15}	962
					$\Delta p/\zeta = -100$				
16.5	2020	3000	134	0.398	0.0912	0.0281	0.621	1.36×10^{-15}	887

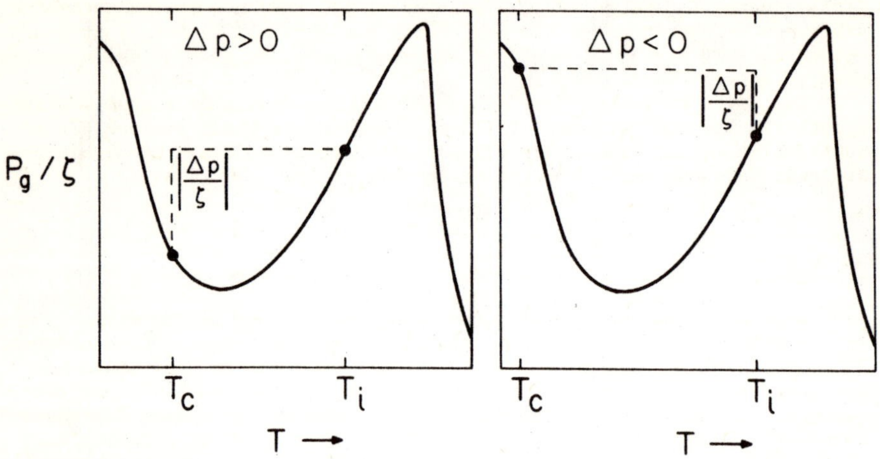

Fig. 4. Schematic diagram illustrating effect of gradients (Δp) in magnetic pressure, turbulent pressure and cosmic ray pressure. If $\Delta p > 0$ the cloud tends to expand; if $\Delta p < 0$ the cloud tends to contract.

(5) Although the results are as yet very uncertain, Jokipii and Lerche (1969) have used a simple formulation of statistical turbulence theory together with available data to show that the interstellar gas has a mean field of 3 μG with rms fluctuations of 1.4 μG and $\langle n_e \rangle \sim 0.04$ cm^{-3} with rms fluctuations of 0.03 cm^{-3}. They discuss a size scale for these fluctuations of the order of 250 pc.

All of these arguments indicate that, independent of the problem of the cold clouds, the inter-cloud medium has accumulations of hot ($T_i \sim 10^3$ °K) matter which probably behave as a system of currents or "turbulence elements."

8. A Possible Sequence for Energization of the Interstellar Medium

There are several interesting implications of the basic model we have assumed for the interstellar gas. The basic assumption is that low energy cosmic rays provide nearly all the thermal energy for the interstellar gas. Furthermore, since the data for the inter-cloud medium seems to indicate that this medium is subject to thermal instability, we would expect it to break up into condensations in a very dynamical manner. This suggests the possibility that these instabilities are responsible for generating the turbulence in the inter-cloud medium. There is then the possibility that the turbulence is responsible for generating magnetic fields. Then, finally, the magnetic fields could pass energy on to the high energy cosmic rays.

It should be noted that the low energy cosmic rays may carry very little energy ($\sim$0.02 eV cm^{-3}, Hjellming, 1969) and have typical stopping distances of the order of 100 pc. Hence the constant re-supply by sources which are very plentiful would be necessary, particularly if these sources have relatively short lifetimes.

The chain of events just discussed has the status of a good working hypothesis for two reasons: (1) Some steps in the chain are subject to qualitative calculations; and (2) Any other possible chain of events has at least an equal lack of evidence to support it.

9. Pulsar Distances

What would be the consequences of an inter-cloud medium with $\langle n_e \rangle = 0.04 \pm 0.03$ cm^{-3}, with a size scale for variations of the order of 250 pc? Any distances obtained by assuming a value for $\langle n_e \rangle$ would be seriously affected. For all pulsars with true distances $\lesssim$ 500 pc, the value of $\langle n_e \rangle$ could vary from 0.01 cm^{-3} to 0.07 cm^{-3}. Pulsar distances based on a constant $\langle n_e \rangle$ would then tend to greatly under-estimate many of the distances and over-estimate others. For this reason the accumulation of pulsars in the vicinity of the sun may be very suspect; they could just be in lines of sight where the value of $\langle n_e \rangle$ takes its lowest values. We would then conclude that unless the $\int n_e ds$ produced by the inter-cloud medium is very large, distances based on an average $\langle n_e \rangle$ may be untrustworthy.

DISCUSSION

R. RAMATY: "Can you make a statement about NP 0527 not being in the Crab since it has a different dispersion measure?"

HJELLMING: "Well, it's off to the side of the Crab. I think that it's unlikely that it is directly associated. Some people put it on the other side of the Crab, some people put it nearer to us, there's really no way one can tell."

L. BAGGERLY: "Since NP 0527 has a lower dispersion measure it is very unlikely that it would be on the far side of the Crab, isn't it?"

HJELLMING: "Well, some people have argued that since there is a large amount of ionized gas in the Crab nebula, contributing to the dispersion of 0532, then 0527 would have a much lower dispersion than 0532 if it were on the near side. Since 0527 has a dispersion measure approximately the same as 0532 it should be on the far side."

S. MARAN: "Don't you think it is now possible to put limits on that by looking at the second order terms in a dispersion now that dispersion at different frequencies is available?"

HJELLMING: "Well, no one has seen any second order terms yet."

T. MATTHEWS: "Since you attribute the dispersion to the interstellar medium, do you subtract away the contribution from the ionization in the Crab nebula itself."

HJELLMING: "No, personally I feel that this contribution from the nebula is probably very small. Of course, that's a gamble that one takes. The whole philosophy here has been that there is a negligible contribution from the region right around the pulsar. If that's not true then most of what I've done is worthless. We really can't do anything except assume that it is all interstellar until some observations prove that that's wrong."

A.M. LENCHEK: "You used your calculations to deduce something about the interstellar cosmic ray spectrum; what did you find out?"

HJELLMING: "The heating of the interstellar gas by cosmic rays depends upon the integral

$$\zeta = \int_{E_o}^{\infty} j(E)\sigma \, dE$$

where j(E) is the cosmic ray differential flux and σ is the ionization cross section. The interstellar flux is related to the observed flux at earth by modulation theory

$$j_E(E) = j_\infty(E)e^{-\eta} .$$

Therefore, what we determine is an integral that depends upon two parameters, E_0 and η. Using the value of η given by Ramaty we find that we get reasonable values for the integral if the cut off energy E_0 is on the order of 0.2 to 1 MeV."

RAMATY: "How much energy is there in the low energy cosmic rays according to this model?"

HJELLMING: "There's very little energy in these low energy cosmic rays. In this model, about 0.02 eV cm^{-3}."

RAMATY: "Another question. Now if the pulsars are really neutron stars and these are the sources of the low energy cosmic rays, then since cosmic rays around 1 MeV have very short ranges, about 10 pc, you would expect very large inhomogeneity for these low energy particles and very non-uniform heating."

HJELLMING: "Well, I would expect the small region in which this kind of particle is being absorbed would show up as a very bright nebula. Now while such a nebula is not present in the Crab nebula, I do think that about one-third of the bright nebulae you see in the sky may be produced by this mechanism."

RAMATY: "But you don't take this into account."

HJELLMING: "Yes I do. The philosophy is that you have to stay away from any line of sight that intersects such a bright nebula. But on the other hand this is a good argument for not applying the value of ζ from one line of sight to another line of sight."

REFERENCES

Bridle, A.H. and Venugopal, V.R. 1969, Nature, 224, 545.
Clark, B.G. 1965, Ap. J., 142, 1398.
Field, G.B., Goldsmith, D.W., and Habing, H.J. 1969, Ap. J. Letters, 155, L149.
Gordon, C.P., Gordon, K.J., and Shalloway, A.M. 1969, Nature, 222, 129.
Gordon, K. J., and Gordon, C.P. 1969 (in preparation).
Hewish, A., Bell, S.J., Pilkington, J.D.H., Scott, P.F., and Collins, R.A. 1968, Nature, 217, 709.
Hjellming, R.M. 1969, Astrophys. Letters (in press).
Hjellming, R.M., Gordon, C.P., and Gordon, K.J. 1969, Astron. Astrophys., 2, 202.
Jokipii, J.R. and Lerche, I. 1969, Ap. J., 157, 1137.
Large, M.I., Vaughan, A.E., and Mills, B.Y. 1968, Nature, 220, 340.
Manchester, R.N., Murray, J.D., and Radhakrishnan, V. 1969, Astrophys. Letters (in press).
Milne, D.K. 1968, Aust. J. Phys., 21, 201.
Rickett, B.J. 1969, Nature, 221, 158.
Rohlfs, K., Mebold, U., and Grewing, M. 1969, Astron. Astrophys. (in press).
Spitzer, L. and Scott, E.H. 1969, Ap. J., 158, 161.
Staelin, D.H. and Reifenstein, III, E.C. 1968, Science, 162, 1481.
Trimble, V. 1968, Astron. J., 73, 535.
Verschuur, G.L. 1969, I.A.U. Symposium No. 39, Crimea (in press).

November 3, 1969

DISPERSION MEASURES AND DISTANCES OF PULSARS

Yervant Terzian

Center for Radiophysics and Space Research
Cornell-Sydney University Astronomy Center
Cornell University, Ithaca, New York 14850

1. Introduction

It is of fundamental importance to astronomers to have a good knowledge about the distances of the celestial objects which they study. In the case of pulsars the frequency dispersion in the arrival time of the pulsed radiation gives an estimate of the total number of electrons in the line of sight to the pulsars. Hence distances to the pulsars can be obtained if we have a knowledge of the electron density distribution in the direction of individual pulsars. Unfortunately, the electron distribution in our galaxy is not well known and the derived pulsar distances using the above method have large errors. To a first approximation one can assume a mean interstellar electron density and derive pulsar distances. However the distances of individual pulsars will have extremely large errors, and from such data it will not be possible to investigate relationships such as the 'period-luminosity' relation of pulsars -- if any exists; except statistically.

The best we can do at present seems to be to investigate all the possible sources of electrons in the direction of individual pulsars separately. Thus one should investigate the HII regions, OB type stars, extended galactic radio sources and other celestial objects which may happen to be in the direction of detected pulsars. From such data one hopes to have a better estimate of the mean electron density in the line of sight to individual pulsars and hence more reliable distances.

In two cases, the Crab and the Vela pulsars, independent distance estimates are available and these can serve as checks on other general methods of distance derivations.

One of the most informative relations in the discussion of pulsar distances is the dispersion measures ($\int n_e ds$) of pulsars and their galactic latitude (b^{II}). Figure 1 shows this relationship for the known pulsars. Clearly the high dispersion measures seem to be concentrated in a band about 10° wide near the plane of the Galaxy; this suggests that these pulsars are considerably more distant than the thickness of the Galactic disk. One can also notice that the equatorial pulsars seem to concentrate a few degrees south of the galactic equator. However statistically this effect does not seem to be very significant.

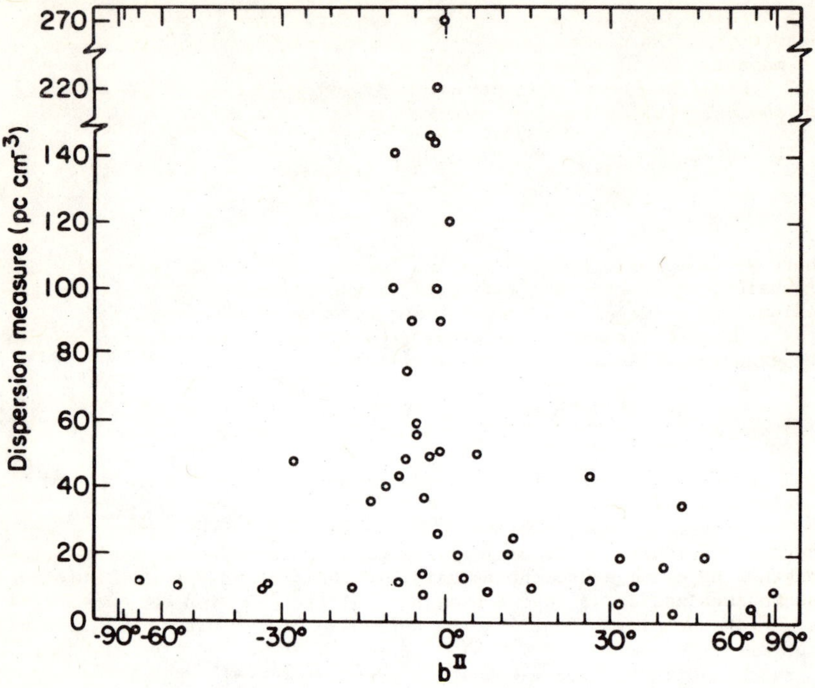

Fig. 1. The relationship between dispersion measure and Galactic latitude.

2. The Equation for Dispersion Measure

The well known Appleton-Hartree equation gives the square root of the complex refractive index of a radio wave in a magneto-ionic medium (see Ratcliffe 1959). For the interstellar medium most of the terms of the Appleton-Hartree equation are negligible and we can use the approximation for the real part of the index of refraction n as

$$n^2 \sim 1 - \frac{X}{1 \pm Y} \tag{1}$$

where
$$X = \frac{e^2 n_e}{4\pi^2 \varepsilon_o m} \frac{1}{\nu^2} = \frac{\nu_p^2}{\nu^2}$$

$$Y = \frac{\mu_o H|e|}{2\pi m} \frac{1}{\nu} = \frac{\nu_H}{\nu}$$

In the above expressions n_e = electron density, e = electron charge, ε_o = permittivity of free space, m = electron mass, ν = frequency of radio wave, ν_p = plasma frequency, μ_o = permeability of free space, H = magnetic field intensity, and ν_H = gyro-frequency.

Taking the square root of n^2 and using the binomial expansion we can approximate the index of refraction as

$$n \sim 1 - \frac{1}{2} X , \tag{2}$$

where we have neglected the 2nd and higher order terms from the binomial expansion, since they are negligible when applied to the interstellar medium.

However a radio pulse travels at the group velocity v_g, so that the group refractive index can be written as

$$n_g = \frac{c}{v_g} = \frac{c}{\frac{d\omega}{dk}} = n + \omega \frac{dn}{d\omega} \tag{3}$$

where c = velocity of light, $\omega = 2\pi\nu$, and $k = 2\pi/\lambda$, λ being the wavelength. By using equation (1) and (2) with (3) and making the same assumptions as for n, one obtains

$$\begin{aligned} n_g &= 1 + \frac{1}{2} X \\ &= 1 + \frac{\nu_p^{\,2}}{2\nu^2} \end{aligned} \tag{4}$$

The change in the group path from its free space value for a radio pulse (caused by electrons along the path s) can be obtained from the difference of the physical path and the group path, and the change in the transit time of the radio wave pulse can then be written as

$$t = \frac{1}{c} \int_o^s (n_g - 1)ds \quad . \tag{5}$$

Substituting from equation (4) into (5) gives

$$t = \frac{1}{2c\nu^2} \int_o^s \nu_p^{\,2} \, ds \quad , \tag{6}$$

and differentiation of equation (6) gives the frequency drift rate of a radio pulse as

$$\left(\frac{d\nu}{dt}\right)_{pulse} = - \frac{c\nu^3}{\int_o^s \nu_p^2 ds} \tag{7}$$

The plasma frequency is a function of the electron density, as can be seen from equation (1), and hence observational information on the frequency drift rates can determine the total number of electrons per square centimeter along the line of sight to a pulsating source. In practice however one uses the time differential in pulse arrival times between two frequencies ν_1 and ν_2, rather than using the frequency drift rates (Drake et al. 1968). This gives greater accuracy. By integrating equation (7) and evaluating the constants for the plasma frequency the "dispersion measure" (DM) in units of cm^{-2} is

$$\int_o^s n_e ds = \frac{2.48 \times 10^4 c(t_{\nu_1} - t_{\nu_2})}{\frac{1}{\nu_1^2} - \frac{1}{\nu_2^2}} \tag{8}$$

where the time delay is in units of seconds of time and the frequencies are in units of MHz. More conveniently we can use the units (cm^{-3} parsec) for DM, thus

$$DM = \int_o^s n_e ds = \frac{2.4 \times 10^{-4} (t_{\nu_1} - t_{\nu_2})}{\frac{1}{\nu_1^2} - \frac{1}{\nu_2^2}} \ . \tag{9}$$

For the pulsar CP 1919, for example, in observations made at the Arecibo Observatory at ν_1 = 73.8 MHz and ν_2 = 111.5 MHz the delay in pulse arrival time at the lower frequency was measured to be 5.328 seconds (Drake et al. 1969). Equation (8) then shows that DM = 3.84×10^{19} cm^{-2} , or about 13 pc cm^{-3}.

3. Methods for Obtaining Pulsar Distances

During the year of 1969 several discussions have been made dealing with the distance determination of pulsars. Rohlfs, Grewing and Mebold (1969) have made use of the 21-cm neutral hydrogen emission observations in the direction of pulsars to estimate pulsar distances. In their analysis they assume that the interstellar medium has two components, a hot gas and a cold one. They then divide the 21-cm line profiles into sections produced by the cold gas ($T = 100°K$) and the hot gas ($T = 5000°K$). Such temperature ranges have recently been proposed by Radhakrishnan and Murray (1969) and others from 21-cm emission observations.

Interstellar electrons are mostly produced by the ionization of the hydrogen gas. A knowledge of the fractional ionization of the hydrogen gas (n_e/n_H) may allow us to compute the electron densities provided that the neutral hydrogen gas density is known. The heating and ionization of the interstellar gas is the result of the low energy cosmic rays and the ultraviolet radiation of the early type stars. Rohlfs et al. (1969) consider only ionizations by low energy cosmic rays, and use the results of Spitzer and Tomasko (1968) for the fractional ionization of the hydrogen gas as a function of temperature. Using the DM equation as

$$DM = \int_o^s n_e ds = \langle n_e \rangle s \tag{10}$$

and assuming that

$$\langle n_e \rangle = \frac{1}{s} \int_o^s n_e ds = \frac{1}{d} \int_o^d n_e ds \tag{11}$$

where d is the path length through the galactic disk in the direction of a pulsar, Rohlfs et al. show that the distance to a pulsar is given as

$$s = d \frac{DM}{\sum \left(\frac{n_e}{n_H}\right)_i N_{Hi}} \tag{12}$$

where N_{Hi} is the hydrogen density of the i component in the 21-cm emission line spectrum, and can be obtained from the 21-cm observations. In order to estimate the path length through the galactic disk Rohlfs et al. adopt a geometrical model of the galaxy adopted from Kerr and Westerhout (1965).

For the Crab and the Vela pulsars their results agree well with the independent optical distance estimates. However, in many cases their derived distances are probably wrong since they have completely neglected the contribution of HII regions to the dispersion measures. Moreover, since they assume that the average of the electron density over the path length through the galactic disk is equal to that over the distance to the pulsar, then if some pulsars lie outside the disk only a lower estimate to their distance can be obtained.

Attempts to determine pulsar distances from the 21-cm line measurements have also been made by de Jager et al. (1968), Guelin et al. (1968), Gordon et al. (1969), Davies (1969), and Bridle and Venugopal (1969). The results of these investigations show that the authors are not in agreement in the derived pulsar distances. However the accepted results from such measurements for the best studied pulsar CP 0329 give a lower limit to its distance of about 1000 pc.

More recently Hjellming et al. (1969) have used the dispersion measure observations of pulsars to determine the properties of the interstellar clouds. These authors conclude that the interstellar hydrogen exists primarily in two forms, one as small clouds with a low temperature of the order of 50°K and

$$(n_e)_c \sim 0.05 \text{ cm}^{-3}, \quad (n_H)_c \sim 10 \text{ cm}^{-3},$$

and the intercloud medium with a temperature of about 1000°K and

$$(n_e)_i \sim 0.03 \text{ cm}^{-3}, \quad (n_H)_i \sim 0.5 \text{ cm}^{-3} .$$

Bridle and Venugopal (1969) determine distances of pulsars, for which there is no obvious evidence of intervening HII regions, by assuming a) that the ratio $\langle n_e \rangle / \langle n_H \rangle = 0.02$ in the galactic disk, and b) by adopting $\langle n_e \rangle = 0.02 \text{ cm}^{-3}$. They argue that the thickness of the galactic disk for electrons is of the order of 800 pc rather than 200 pc, which is the thickness obtained from the neutral hydrogen distribution. The n_e galactic thickness results from the 10 MHz continuum observations of the galaxy.

Davidson and Terzian (1969a) have also discussed the distance determination of pulsars by adopting several methods applicable to individual cases. They discuss the DM versus $|b^{II}|$ diagram and estimate electron densities; some pulsars are assumed to fit the spiral arms of the galaxy; and for some cases $\langle n_e \rangle = 0.03 \text{ cm}^{-3}$ is assumed, which represents the average electron density from the derived distances. These authors also indicate that the contribution of the HII regions in the line of sight to some pulsars contribute significantly to the dispersion measures.

Perhaps the most detailed analysis for pulsar distances is that made by Prentice and ter Haar (1969). Their study includes computations of Strömgren spheres for all O-type and supergiant stars within 1 kpc from the sun. Thus they estimate the dispersion measures that result from the ionization of the neutral hydrogen by the early type stars. They also discuss the ionization due to cosmic rays. By combining their results with the observed dispersion measures they compute pulsar distances. Their results indicate that within 1 kpc from the sun the median value for the interstellar electron density is about 0.04 cm^{-3}, in good agreement with the results of Davidson and Terzian of 0.03 cm^{-3}.

4. HII Regions and Dispersion Measures

Since pulsar distances can be estimated from the observed dispersion measures one should carefully examine the contribution of the HII regions to the dispersion measures.

The energy for the ionization of the neutral gas in an HII region is provided by the early type OB and super-giant stars. The number of Lyman continuum photons emitted by such stars is reduced in an HII region due to the re-combinations taking place to excited levels of the hydrogen atoms. We thus can define the radius of a Strömgren sphere s_o as the radius of an HII region where at its boundary the net ultraviolet flux streaming outward is zero, i.e., at the boundary the number of Lyman continuum quanta are balanced by the recombinations in the ionized volume. Hence we can write for a pure hydrogen cloud

$$4\pi R^2 N_{Lc} = \frac{4}{3}\pi s_o^3 n_e^2 \alpha(T) \qquad (13)$$

where R = radius of the exciting star, N_{Lc} = number of Lyman continuum quanta cm^{-2} sec^{-1}, and $\alpha(T)$ = recombination coefficient (2.6×10^{-13} cm^3 sec^{-1} at $T = 10^4$ °K). Solving the last expression for s_o,

$$s_o = \left[\frac{3R^2 N_{Lc}}{n_e^2 \alpha(T)}\right]^{1/3} \qquad (14)$$

Theoretical s_o values as a function of the spectral type of the early type stars have been computed by several authors (Spitzer 1968, Rubin 1967). The Strömgren sphere radii as a function of the early type stars for $n_e = 1$ cm^{-3}, $T = 10^4$ °K as given by Spitzer are shown in Table 1. It is evident that if in the line of sight to a pulsar one or more early type stars exist, then we would expect the dispersion measure for that pulsar to be somewhat higher than the dispersion measure for one where no OB type stars are present in the line of sight to it.

If we consider the well known large HII regions in the galaxy we find that the dispersion measure of a pulsar which may happen to be located behind such a large HII region will be increased to extremely high values. Table 2 shows the dispersion measure contributions from several well known HII regions. Dispersion measure values of the order of 500 to 1000 pc cm^{-3} are indeed expected from many well known HII regions. However with presently available instrumentation the highest detected value for the dispersion measure of a pulsar is 270 pc cm^{-3}.

There are many OB and supergiant type stars in the galaxy which are not directly associated with dense HII regions. Therefore the Lyman continuum radiation from such stars will mostly contribute to the ionization of the general interstellar medium, which has a mean electron density of the order of 0.03 cm^{-3}. The Strömgren sphere radius of an 05 type star in such a medium will be about 1000 pc.

TABLE 1

STRÖMGREN SPHERE RADII

(n_e = 1 cm^{-3}, T = 10^4 °K)

Spectral Type	S_o (pc)
05	100
06	66
07	44
08	29
09	18
B0	11
B1	6.4

TABLE 2

DISPERSION MEASURES DUE TO HII REGIONS

HII Region	Diameter (pc)	$\langle n_e \rangle$ (cm^{-3})	DM (pc cm^{-3})
Orion	0.83	1.65×10^3	1370
NGC 2024	0.69	7.5×10^2	510
M 16	9.31	8.7×10^1	810

Studies of the dispersion measure of pulsars and HII regions have been made by Davidson and Terzian (1969b), and as mentioned earlier, by Prentice and ter Haar (1969). Davidson and Terzian have used the luminosity function for early type stars in the galaxy together with the theoretical Strömgren sphere radii for such stars, and have computed the line-of-sight mean free path which corresponds to the distance which one may expect to travel along a straight line in the galactic plane before encountering an HII region. Their computations show that the line of sight to a pulsar at a distance of 2 to 3 kpc will encounter at least one HII region associated with the Strömgren sphere of an early type star.

5. The Dispersion Measure of the Crab Pulsar

The dispersion measure of the Crab nebula pulsar, NP 0531, is 57 pc cm^{-3}. Since the Crab nebula is at a distance of 2 kpc, then $\langle n_e \rangle \simeq 0.03$ cm^{-3}. The Crab nebula is located in the galactic anti-center and is a member of the Perseus spiral arm. Hence the line of sight to NP 0531 passes through an inter-spiral arm region, where one may expect that the mean electron density may be lower than the one in a spiral arm.

The neutral hydrogen in the 2 kpc line of sight toward NP 0531 is about $\int n_H ds \simeq 1000$ pc cm^{-3} (Hjellming et al. 1969); this implies that the mean fractional ionization $\langle n_e \rangle / \langle n_H \rangle \simeq 5\%$. This result is somewhat higher than what one expects if the ionization is due to low energy cosmic ray particles. However the ionized gas within and around the Crab nebula may account for part of the observed dispersion measure. Woltjer (1958) has suggested that the Crab nebula may contain a diffuse mass with a density $n_e \simeq 40$ cm^{-3}. If the mass of the nebula is $\sim 0.3 M_\odot$ then it would occupy a sphere of radius 0.5 pc, and thus would contribute 20 pc cm^{-3} to the dispersion measure of NP 0531 (provided the pulsar is in the center of the nebula; probably a good assumption).

Furthermore, Davidson and Terzian (1969a) have shown that the interstellar gas surrounding the nebula must be ionized by the ultra-violet radiation of the supernova and its remnant. These authors estimate the Strömgren sphere radius appropriate to the present ionizing luminosity (and age) of the Crab nebula and find that the resulting ionized sphere would have a radius of about 80 pc, thus contributing about 12 pc cm^{-3} to the dispersion measure of NP 0531. Hence the observed dispersion measure of NP 0531 is the sum of the dispersion measure contributions from the Crab nebula, the Strömgren sphere of the Crab nebula, and the general interstellar medium. The above data indicates that the dispersion measure due to the general interstellar medium is 24 pc cm^{-3}, which at the distance of NP 0531 gives $\langle n_e \rangle \simeq$ 0.012 cm^{-3}. This somewhat low density value should not be surprising since, as we mentioned above, the line of sight to the Crab passes mostly through an inter-spiral arm medium where low densities are expected.

6. Pulsar Distances

In this paper we have attempted to present a set of pulsar distances taking into account all known possible effects which may contribute to the observed dispersion measures, as discussed in the previous sections. Table 3 lists the presently known 49 pulsars (December 31, 1969) and gives their designations, galactic coordinates, dispersion measures, distances, and mean electron densities in the direction of each pulsar. The letter symbols after the reported distances in Table 3 have the following meanings; a: $\langle n_e \rangle = 0.03$ cm^{-3} assumed; b: distance assumed to fit spiral arms; c: distance from optical identification; d: 21 cm line absorption; e: assumed that z (galactic half-width) = 200 pc (for $b^{II} > 35°$); and f: $\langle n_e \rangle = 0.10$ cm^{-3} assumed for small b^{II}, and presence of HII. The last column in Table 3 gives some remarks relevant to the dispersion measures and distances of the pulsars. These remarks obviously help in determining the best applicable method in estimating

TABLE 3

PULSAR DISTANCES AND DISPERSION MEASURES

Designation	Pulsar	ℓ^{II} (o)	b^{II} (o)	$\int n_e ds$ (pc cm^{-3})	Distance (pc)	$\langle n_e \rangle$ (cm^{-3})	Remarks
P0031-07	MP 0031	111	-69	12	210 e	0.06	
P0254-54	MP 0254	271	-55	10	240 e	0.04	
P0329+54	CP 0329	145	- 1	27	2000 bd	0.01	Lower limit of 1 kpc from 21 cm absorption.
P0450-18	MP 0450	217	-34	10	330 a	0.03	Just misses Strömgren sphere of I Ori.
P0525+21	NP 0527	184	- 7	49	2000 b	0.02	DM about the same as Crab pulsar.
P0531+21	NP 0531	184	- 5	57	2000 c	0.03	Crab pulsar.
P0628-28	MP 0628	237	-17	10	330 a	0.03	
P0736-40	MP 0736	254	- 9	100	1000 f	0.10	In the direction of HII ξ Pup and γ^2 Vel.
P0808+74	CP 0808	140	32	6	190 a	0.03	
P0818-15	MP 0818	237	12	25	830 a	0.03	
P0823+26	AP 0823+26	197	32	19	630 a	0.03	
P0833-45	PSR 0833-45	264	- 3	50	500 c	0.10	Vela X pulsar.
P0834+06	CP 0834	220	26	13	430 a	0.03	
P0835-40	MP 0835	260	1	120	1200 f	0.10	In the direction of HII ξ Pup and γ^2 Vel.
P0904+77	PSR 0904+77	135	34	---	---	---	
P0940-56	MP 0940	279	- 3	145	2300 b	0.06	In the direction of HII γ^2 Vel.
P0943+08	PP 0943	228	42	18	300 e	0.06	

TABLE 3

PULSAR DISTANCES AND DISPERSION MEASURES (Continued)

Designation	Pulsar	ℓ^{II} (o)	b^{II} (o)	$\int n_e ds$ (pc cm^{-3})	Distance (pc)	$\langle n_e \rangle$ (cm^{-3})	Remarks
P0950+08	CP 0950	229	44	3	290 e	0.01	
P0959-56	MP 0959	281	- 1	90	2300 b	0.04	Just misses Strömgren sphere of γ^2 Vel and IC 2602.
P1133+16	CP 1133	242	69	5	210 e	0.02	
P1154-62	MP 1154	297	0	270	2700 f	0.10	In the direction of HII α and β Cru.
P1237+25	AP 1237+25	253	87	9	200 e	0.05	
P1240-64	MP 1240	302	- 1	220	2200 f	0.10	In the direction of HII α and β Cru.
P1359-50	MP 1359	314	11	20	670 a	0.03	
P1426-66	MP 1426	313	- 5	60	1500 b	0.04	In the direction of HII β Cru.
P1449-65	MP 1449	315	- 5	90	1700 b	0.05	Just misses Strömgren spheres of δ Cir and HD 135591.
P1451-68	MP 1451	314	- 8	12	400 a	0.03	
P1507+55	MP 1507	91	52	20	250 e	0.08	
P1530-53	MP 1530	326	2	20	670 a	0.03	
P1541+09	AP 1541+09	18	46	36	$\leq$1200 a	$\geq$0.03	In the direction of the North Galactic Spur.
P1604-03	MP 1604	8	34	10	330 a	0.03	

TABLE 3

PULSAR DISTANCES AND DISPERSION MEASURES (Continued)

Designation	Pulsar	ℓ^{II} (o)	b^{II} (o)	$\int n_e ds$ (pc cm^{-3})	Distance (pc)	$\langle n_e \rangle$ (cm^{-3})	Remarks
P1642-03	PSR 1642-03	14	26	40	1300 a	0.03	Close to the North Galactic Spur and HII ζ Oph.
P1706-15	MP 1706	7	15	10	330 a	0.03	
P1727-50	MP 1727	341	- 9	140	1500 b	0.09	Close to the boundaries of HII I Scorp.
P1747-48	MP 1747	344	-11	40	1500 b	0.03	Close to the boundaries of HII I Scorp.
P1749-28	PSR 1749-28	2	- 1	51	1600 b	0.03	Region of moderate obscuration.
P1818-03	MP 1818	27	6	50	2400 b	0.02	
P1857-25	MP 1857	11	-13	35	1200 a	0.03	
P1911-05	MP 1911	31	- 7	75	750 f	0.10	Region of high obscuration.
P1919+21	CP 1919	56	4	13	420 a	0.03	
P1929+10	PSR 1929+10	47	- 4	8	270 a	0.03	
P1933+16	JP 1933+16	52	- 2	143	1430 f	0.10	Region of high obscuration.
P1944+17	MP 1944	54	- 4	35	≤1200 a	≥0.03	Close to HII S84.
P2015+28	AP 2015+28	68	- 4	14	470 a	0.03	
P2022+51	JP 2022	88	8	10	330 a	0.03	
P2045-16	PSR 2045-16	31	-33	11	380 a	0.03	
P2113+46	JP 2113	89	- 1	100	1000 f	0.10	Region of high obscuration.
P2218+47	PSR 2218+47	98	- 8	44	2500 b	0.02	
P2303+30	AP 2303+30	98	-27	46	460 f	0.1	Close to extended nebula.

the distance to individual pulsars. The dispersion measure for the pulsar P 0904 + 77 has not yet been reported in the literature and hence no distance estimate has been made for it.

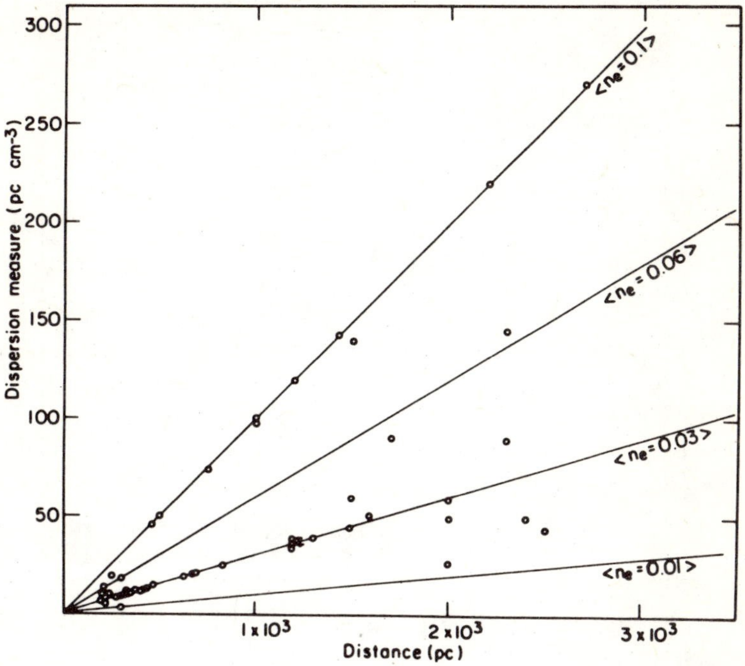

Fig. 2. Pulsar distances and dispersion measures.

The results of this analysis shows that pulsars are at distances from a few hundred to a few thousand parsecs away and that the mean interstellar electron density within a few kiloparsecs from the sun varies from $\sim$0.01 to $\sim$0.10 cm^{-3}. Figure 2 shows a plot of the pulsar distances with dispersion measure. Figure 3 shows the distribution of pulsars projected on the galactic plane, where the distances have been taken from Table 3.

The mean electron density from all the data in Table 3 is $\sim$0.045 cm^{-3}. A pulsar at a distance of 10 kpc with a mean interstellar electron density in its line of sight of 0.05 cm^{-3} will have a dispersion measure of 500 pc cm^{-3}. The presently available instrumentation may not be able to detect such a pulsar with such a high dispersion measure.

7. Pulsars in Other Galaxies

Of great cosmological importance is the knowledge of the intergalactic density. There are almost no direct observational results dealing with the intergalactic matter. At first glance one could see that detection of pulsars in other galaxies may give direct observational

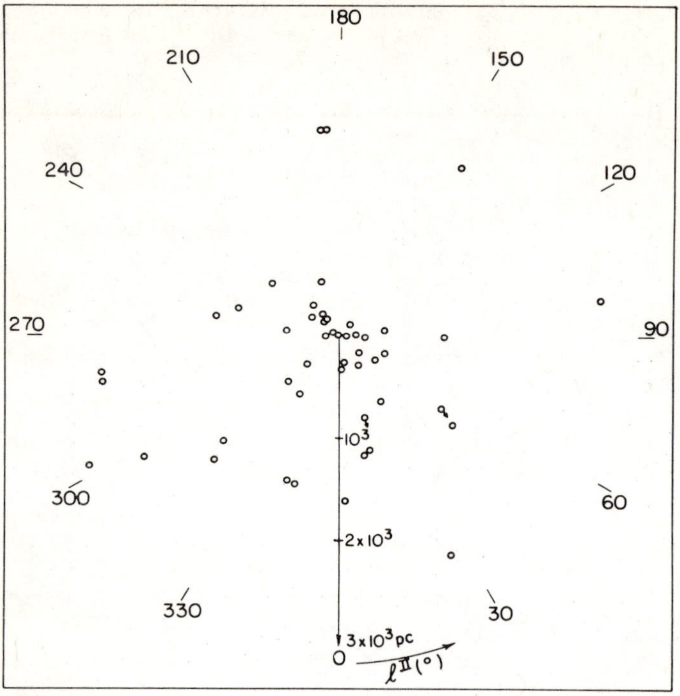

Fig. 3. The distribution of pulsars projected on the Galactic plane, using data from Table 3.

detection of the intergalactic matter by its contribution to the dispersion measures of such pulsars. However due to the extremely low cosmical densities the dispersion measure contribution by intergalactic matter compared to the galactic one seems to be very small, for the nearby galaxies.

Taking an upper limit for the intergalactic density of 10^{-28} gm cm^{-3}, we have $\langle n_e \rangle \simeq 0.0001$ cm^{-3}. At the distance of the Andromeda galaxy M31, d = 570 kpc, the dispersion measure contribution due to the intergalactic electrons will only be about 60 pc cm^{-3}. Clearly then the bulk of the dispersion is from the galactic interstellar matter.

ACKNOWLEDGMENT

This work was supported by the Arecibo Observatory. The Arecibo Observatory is operated by Cornell University under contract to the National Science Foundation and with partial support from the Advanced Research Projects Agency.

REFERENCES

Bridle, A.H., and Venugopal, V.R. 1969, Nature, 224, 545.
DeJager, G., Lyne, A.G., Pointon, L., and Ponsonby, J.E.B. 1968, Nature, 220, 128.
Davidson, K., and Terzian, J. 1969a, A.J., 74, 849.
Davidson, K., and Terzian, Y. 1969b, Nature, 221, 729.
Davies, R.D. 1969, Nature, 223, 355.
Drake, F.D., Gundermann, E.J., Jauncey, D.L., Comella, J.M., Zeissig, G.A., and Craft, H.D. 1968, Science, 160, 503.
Gordon, C.P., Gordon, K.J., and Shalloway, A.M. 1969, Nature, 222, 129.
Guélin, M., Guibert, J., Huchtmeier, W., and Weliachow, L. 1968, Nature, 221, 249.
Hjellming, R.M., Gordon, C.P., and Gordon, K.J. 1969, Astron. Astrophysics., 2, 202.
Kerr, F.J., and Westerhout, G. 1965, Galactic Structure, ed. A. Blaauw and M. Schmidt, Univ. of Chicago Press, Chicago.
Prentice, A.J.R., and ter Haar, D. 1969, M.N.R.A.S., (in press).
Radhakrishnan, V., and Murray, J. 1969, Proc. Astr. Soc. Aust., (in press).
Ratcliffe, J.A. 1959, The Magneto-Ionic Theory and its Applications to the Ionosphere, Cambridge University Press, London.
Rohlfs, K., Grewing, M., and Mebold, U. 1969, Forschungsbrichte Der Astronomischen Institute, Bonn, No. 69-03.
Spitzer, L. Jr. 1968, Diffuse Matter in Space. Interscience Publ., New York.
Spitzer, L. Jr., and Tomasko, M.G. 1968, Ap.J., 152, 971.
Woltjer, L. 1958, B.A.N., 14, 39.

January 1970

ROTATING NEUTRON STARS AND THE NATURE OF PULSARS

Thomas Gold

Cornell University
Ithaca, New York

At the time of the discovery of the pulsars there were several considerations, those which are at the forefront now, that led me to suggest that the pulsars were rotating neutron stars. These are: the enormity of the energy density being radiated, the fact that the energy content of rotation of a collapsed star of neutron star dimensions is an enormous quantity and the fact that magnetic field strengths of a collapsed star would reach very high values. These kinds of considerations immediately make one think that a rotating magnetic field is a much better machine for making instense pulses than something involving vibration. Rotation has a systematic nature that vibration does not have. Violent pulsation is likely to destroy the transmitter every time a pulse is fired while rotation is a very much more organized affair.

The mass numbers of stable nuclei range from 1 to 238, then there is a gap and it begins again at about 10^{56}. There is a very small range near 10^{56}, or about one solar mass, where a stable nucleus, the neutron star, is possible. Following this range there is then a semi-infinite range of stability, the Schwarzchild singularities. The stable range that includes the neutron stars is very small, from about 0.5 $M_\odot$ to about 2 $M_\odot$. It is funny that this range is so tiny. It would have been so easy for the constants of nature to have been slightly different so that the onset of the Schwarzchild singularities came before the neutron star range, but it doesn't.

If one studies the mass-radius relations of various astronomical objects (Figure 1), then there is a small part of the mass versus radius diagram in which an object can resist the gravitational force by means of internal <u>non-thermal</u> pressures (K. Thorne has tried to calculate the shape of this domain). This domain has two curves of stable objects, the white dwarfs and the neutron stars. One can think of all the objects in the universe sitting around debating how they are going to make it from the ordinary physics part of the diagram to the Schwarzchild singularity part, because that is what they are trying to do. Ordinary stars have to sit and radiate away their energy, which takes some long time, on the order of 10^{10} years, before they can reduce their radii. Since there are no stable configurations in between, they can then only collapse discontinuously. If their masses are small enough they can stop their collapse at the white dwarf or neutron star range. If their masses are in the fortunate range, from $\sim$2 to $\sim$50 $M_\odot$ they will not stop but will jump right to the singularity stage. These are explosive events for the outer part, but a collapse for the inner regions. At still greater

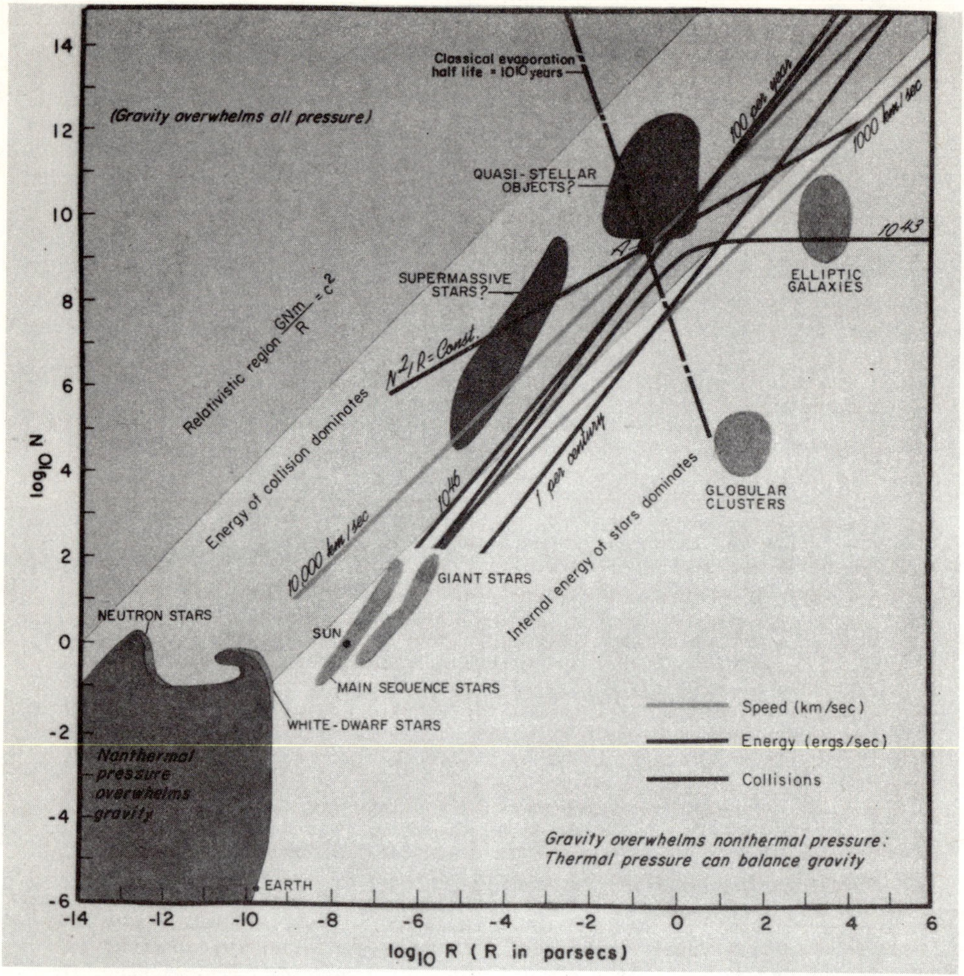

Fig. 1. Diagram of log N against log R (N is the mass of the astronomical object divided by the mass of sun, R is the characteristic radius). All known astrophysical objects outside the solar system, with the exception of white dwarfs and neutron stars, lie in the region where gravity overwhelms non-thermal pressure (such as the pressure of electrons in atoms) and can be balanced by thermal pressure.

The line N^2/R = const is a line along which a cluster will evolve by evaporation of stars that take away no more than the energy of escape.

The energy release shown is that due to geometrical collisions giving up the kinetic energy, plus the luminosity of all the stars. Collision rate is for geometrical collisions only.

This figure is a composite of diagrams by T. Gold (High Energy Astrophysics Vol. 1, p. 199, Gordon and Breach, N.Y., 1967.) and by K.S. Thorne (Scientific American, 217, 88, 1967.)

masses there are no stable stars to start out with, but rather clusters of stars. These then have an additional form of energy, the Virial energy, to get rid of before they can collapse, and that takes a long time. For them to get rid of their internal energy finally, they have to go through a regime of star collisions. The cluster may collapse to a superstar that can jump to the singularity state, but before it gets there it has to go through a stage in which the collision speeds are very high, maybe 10^4 km sec^{-1}. Quite a lot of astronomy must be taking place in that domain and it is a pity we don't have a few of these regions close enough to see. But if you have the general point of view that things are always progressing toward the Schwarzchild singularity domain, then we ought to have places where star collisions are occurring at ten thousand kilometers a second and that's quite some collision. So there ought to be singularities of masses of say 10^9 $M_\odot$. Are these the quasars? Or are the star collision regions the quasars? Or are there huge singularities around in which there is material orbiting at relativistic speeds? I don't know but it's fun to discuss astronomy from the point of view of this mass-radius diagram.

Now, the neutron star structure has been studied by Ruderman (1968), Salpeter (1961) and others. The central density is $\sim 10^{14}$ g cm^{-3}, quite a respectable density. There is an atmosphere which, with an acceleration of gravity $\sim 10^{12}$ g (g is the gravitational acceleration on the earth), is only a few centimeters high. This is followed by a region of white dwarf matter, then a region of neutron-rich soup containing nuclei (mostly iron). In this region containing nuclei there will be a crystalline structure. Since the temperatures are well below the melting point this will be a good solid. It is important to discuss this region because it implies that rotation will be rigid-body rotation (Figure 2).

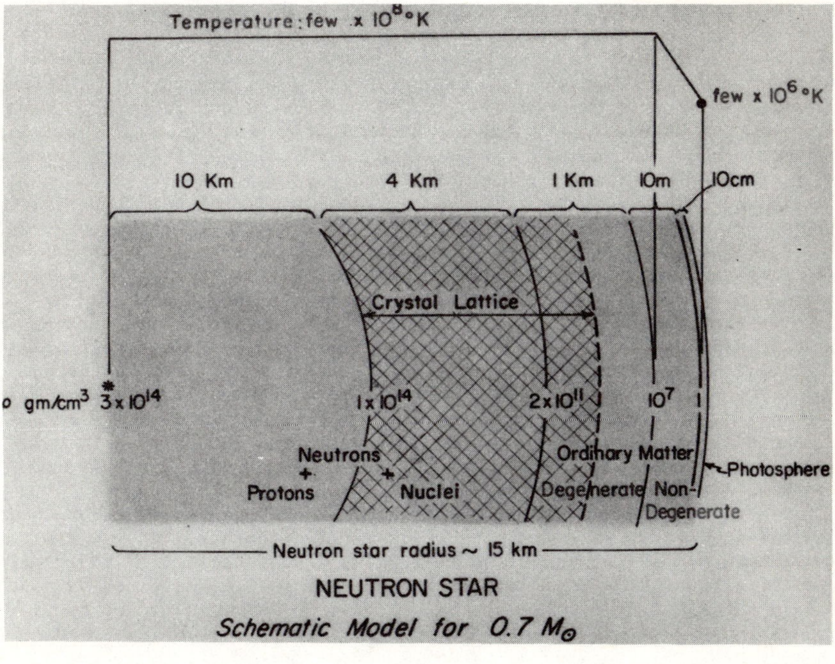

Fig. 2. Schematic model for a neutron star as suggested by E.E. Salpeter.

If one plots pulse length versus period of the known pulsars, (Figure 3) there is some scatter but there is enough regularity to conclude that they are not dissimilar objects. They are all reasonably similar with the chief variable being rotation speed. This is not a strong argument, but at any rate the observations are compatible with the idea of similar objects rotating at different speeds.

Regarding distances, we know the distances to some pulsars (the Crab, Vela approximately, one or two others from 21 cm absorption). The distribution of pulsars versus galactic latitude gives some information about their distances. If they were very close compared to the thickness of the galactic sheet there would have been no concentration near $b^{II} = 0$ (b^{II} = galactic latitude) (as there is) or if they were all very far away they would all be seen by us as very close to the plane of the Galaxy (but a large fraction are at high latitude). So they are neither very close nor very far and if one makes a Monte Carlo calculation with the stellar distribution to reproduce the observed latitude distribution then that tells us that their distance apart is $\sim$500 pc. This was learned approximately as soon as five or six pulsars were known.

The distance is important for the theory of the emission process because it bears on the question of the total energy radiation rate. Since that estimate comes to such enormous values it obviously is very important.

Knowledge of the approximate distances and the observed radio fluxes gives one an idea of how much power is being emitted. From what volume? From a volume that must be small enough to give as short a pulse as is observed, so that the time-of-light-travel across the emitting region must be no longer than the pulse length involved. In many cases we are talking about fractions of milliseconds for pulses. This means small emitting regions. Of course, in any one case you could invent a huge emitting sheet oriented so that it is just facing the earth, but that is a very artificial situation. Then it would be an organized affair, to be facing just the earth, and that would only be so if these indeed were the Little Green Men that Cambridge originally thought caused pulsars. But if you take it that the pulse length really defines the size of the emitting region then you come to some very large values for the energy flux. If the emission were due to independently emitting particles of electronic charge, then the system would be subject to the thermodynamic consideration that the energy of each particle would then have to correspond to the temperature at which a black body would radiate the required power in the radio band. One then finds that the particles would need to have energies of 10^{21} eV and that really seems unlikely.

Therefore you must say that you are seeing some correlated radiation process in which many charges travel together. In most cases you are thus talking of 10^8 to 10^{12} charges moving in synchronism in emitting the radio band radiation: a highly organized affair. Now some people talk about plasma instabilities that will assure that a lot of charges move in step. It may be so, but I have not yet been able to be satisfied that anybody understands mechanisms that have that degree of perfection. It is necessary for that sort of mechanism to be remarkably perfect to have a brightness temperature of 10^{25} °K in the radio band and not to have some little, subsidiary processes go and dump all the energy into hard X-rays rather than in the radio band. That is very remarkable. You are dealing with a transmitter that doesn't get too hot. If you let a very little bit of gas into your radio tube then instead of making radio waves it will make a lot of higher energy quanta -- it will get hot and radiate infrared. The pulsar mechanism must be very clean. It is true that in the case of the Crab the spectrum goes up to the X-ray band, but

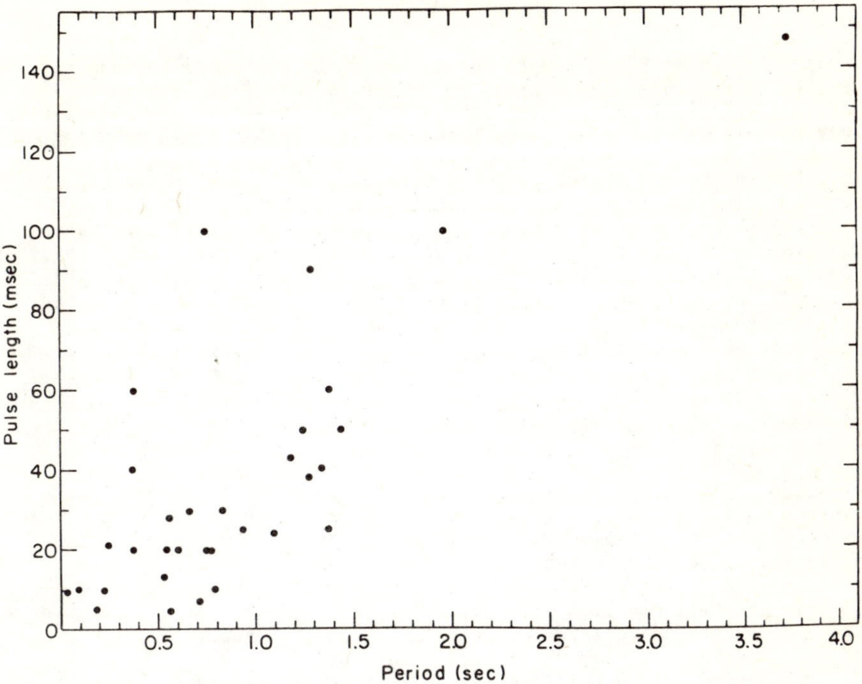

Fig. 3. Pulse length vs. pulse period for 32 pulsars.

it does not dump all its energy into the high frequencies and leave nothing for the low frequencies. In most cases the spectrum rises toward low frequencies which is very remarkable for a body that emits energy densities that are very high. It must have no other degrees of freedom into which it can put its energy because all the higher modes, by their greater number, would then make the ultraviolet catastrophe and make all the radiation come out at the high frequencies. It is not so, therefore it must be a highly organized system.

If one examines the pulse shapes in successive cycles of a given pulsar, one sees that the pulses wobble around a bit but do not show any systematic drift. So there is some phase jitter but there is terribly little frequency jitter. Only when you look with very high accuracy, as we are now doing with the Crab, do you see any progressive drift of pulses compared with the predicted position, but then only of the order of one part in 10^9.

The pulses are frequently double. Now, I think it is bad tactics to try to account for the details of the pulse shapes. I recall that when seven pulsars were known some investigators had theories for accounting for seven pulse shapes. Now they would need 44 theories and I doubt that they are keeping up. There is nothing very systematic about the pulse shapes. Different pulsars do in fact have different characteristic pulse shapes but among those the double humped shape is remarkably common. This is certainly a significant thing.

Many pulsars have a lot of linear polarization and show a characteristic motion of the polarization vector throughout the pulse. This is another well-organized effect and must be considered in any theory.

Another well-organized effect is the march of the subpulses from the back to the front of the pulse. It occurs in several pulsars. This augmentation of the pulse does not happen outside the pulse. I do not know what it means, but it does imply that the system is in some way very systematic.

I find it much easier to account for all these systematic effects with a rotating object than a pulsating one.

The pulse length as a function of frequency is a very important function, I think. The pulse length does not change very much with frequency. Now in a rotating model we are talking about something like the beacon of a lighthouse. How come, we would say, that that beacon has almost the same beam-width over a huge range of frequency? Now if I design an antenna, it is very difficult to make its beam-width independent of frequency. So you might say, surely that argues against a rotating model. But, on the other hand, think what it would be for any other kind of system. For any pulsating system, it would mean that the different frequencies would have to be emitted in the same place in space, which is very unlikely. But the rotating antenna problem has a very good solution, one that I find very palatable. If the radiation is emitted by a certain sector around the star, and if the radiation is due to particles whose motion is sufficiently relativistic, then we can get just this phenomenon. Because if the particles, singly or collectively, have a relativistic factor γ then the pulse length will be proportional to $1/\gamma$. If this is small compared with the width of the sector around the star whe where there are charges radiating to us, then this beacon has the quality of not changing its beam-width with frequency.

I want to soften this argument a little bit because there is some slight variation of pulse width with frequency. If there is a distribution of gammas, as we would expect, then the lower energy particles would contribute more to the lower frequencies and would make the pulse width a little wider at low frequency.

In discussing the radiation mechanism, there are three basic points that I understand. One is that we have a rotating neutron star with a fieldstrength on its surface that is given by the collapse of the previously existing field and the law of conservation of flux. The fields may then be $\sim 10^{12}$ gauss. The kinetic energy of rotation is very substantial, around 10^{52} ergs initially. This is of the same order as the total nuclear energy that the star possessed before it collapsed. So the star shines for maybe 10^9 to 10^{10} years on its nuclear energy. Then it disastrously collapses and winds up with as much energy in its spin as it used up in its previous incarnation.

Now there are three different points of view as to what might be the dominant situation concerning the details. One is Goldreich's view that there must be a homopolar dynamo in such a situation (Goldreich and Julian, 1969). This must be so for the pulsars because, whatever the corotation situation may be close to the star, there is no question that far from the star matter is not corotating. Then there must be a current or if there is no current there must be an emf which turns out to be 10^{18} or 10^{19} volts. Since we do not believe that there is enough polyethylene around a pulsar to hold off that many volts, the current must flow. It is no good to say there may be a vacuum; a vacuum doesn't hold off volts. It is the walls of the vacuum that have to hold the voltage and to be able to prevent current flowing when there is a voltage of 10^{18} volts requires an insulation that we certainly don't see in any astronomical case.

In this current I do not see any way of getting the very high energy particles ($\sim 10^{13}$ eV) that are needed to keep the Crab shining. Also no one has shown how the homopolar dynamo can result in a radiation mechanism of the intensity I have described. But the current must be there and I assume that there is some gas throughout the region out to the light cylinder (where corotation implies the speed of light). Just what happens to this plasma is not quite clear to me.

The second phenomenon that will go on is that out to a distance close to the light cylinder radius, or critical distance, corotation will be enforced. There will be the appropriate combination of radial electric field and magnetic field to ensure corotation, just as in the earth's magnetosphere. Close to the critical radius there is the relativistic increase in mass of the particles and one is no longer quite sure what happens.

Now I do not understand all the motions that would be set up in this corotation. Perhaps charges would be flung out to the vicinity of the critical radius, to a speed close to the speed of light. Perhaps the particles would reach a magnetic rigidity of just the value that could be contained by the fieldstrength at the critical radius.

The third discussion is whether the radiation and the energetic particles are both the result of processes that go on outside the critical radius and that are the consequence of merely radiating the magnetic dipole radiation due to a skew magnetic field. That is the Pacini (1968), Ostriker-Gunn (1969) discussion. Personally, I would be surprised if the main acceleration process went on far outside the critical radius because I feel you are diluting everything and yet you are trying to make a theory that makes these enormously intense, concentrated sources of radiation. Putting in the low frequency wave as an intermediary, and having that pick up particles in turn, seems likely to diffuse the mechanism into bigger and bigger volumes. Also I would argue against it on various other grounds. I can't see how that is likely to make a pulse length that is independent of frequency, for example, because that is something that is marching outwards and there are going to be different levels where preferred frequencies are emitted. It is very unlikely that all frequencies will be emitted at the same level.

For the case that I think is most likely I want to give a little more detail. Suppose the neutron star has a non-axial dipole field. Now we all know about the way particles move in the magnetic "bananas" in the magnetosphere. What would happen to the banana whose outer edge is going around at a relativistic speed? I think that what it will do is make energetic particles which will have a significant motion parallel to the magnetic field. Their radiation will come from various parts of the tube of force. As you look at it you only see the radiation from particles that are momentarily moving toward you. In general there will be two places on the banana where this condition holds. If the banana is not precisely in the meridional plane, these two places will not be superimposed at you in time. Typical magnetospheric motions thus seem to have some element in the right direction toward explaining the double pulses.

Now I remember what a hard time one had to find out what was going on in the earth's magnetosphere, at low energies, with rotation an insignificant factor. It is much harder still when compounded with the notion that there is a relativistic rotation speed in the whole picture. Would you then really be able to say: "I understand exhaustively all the processes taking place?" It is certainly possible to make high energy particles in such a situation, but would you be able to say now where and how they would be produced? I find that very difficult to know at this stage. It is such an elaborate problem that I do not think that one can make up one's mind right here and now just what would be taking place. You certainly have a situation where the energy of rotation is available for making fast particles.

One more point. I said one needs of the order of 10^8 charges radiating together. How can that happen? I believe the following problem has to be solved to answer this question. If one has an ordinary synchrotron and it radiates, that is in practice due to independent charges. But suppose I were to put many more electrons into that same synchrotron so that they are very close together. Each electron that wants to radiate would feel out whether it has any neighbors that are moving in such a way that their radiation would be in phase with the first one. If there are such neighbors they will push on each other. Will they drive each other into bunches or apart? We know that the first order effects come to nothing but the second order effects we don't know. I think this would be very important to solve. If the radiation reaction indeed augments the bunching, then I feel that is the answer. Many processes will tend to make high energy particles in the rotating magnetosphere; then their own radiation would drive them into bunches and produce the highly organized system one needs. If the answer turns out to be that the radiation reaction tends to drive the particles apart then I would really feel that the whole pulsar story is still a very deep mystery.

DISCUSSION

R. RAMATY: "Regarding your comment on the magnetic dipole radiation of Pacini and Ostriker, you gave the impression that that mechanism is somehow not good. Do I understand correctly you are saying that it cannot be responsible for the energy loss?"

GOLD: "Well, it can be responsible for the energy loss but it is not true to say that you must have the low frequency radiation, for several reasons. One is that it might be greatly diminished by the fact that if the neutron star can accommodate its shape to hydrostatic equilibrium, as seems likely, then the skew dipole will align itself. The dipole field has a slight tendency to flatten the star in the equatorial plane of the field. If this is not coincident with the equatorial plane of rotation it will make a force that will tip the magnetic axis towards the rotation axis. The time constant for this is very short. If the field is more complex, then you have nevertheless greatly diminished the whole effect if you are left with only the higher modes of the field that you can't get rid of by this mechanism. The other thing is that the low frequency radiation, which is way below the plasma frequency, is essentially going to fill a certain sphere around the star but will not propagate. To equate an energy loss with a non-propagating wave is a very dubious thing. The production and radiation from the energetic particles may be the chief energy loss process. I think any one pulsar will radiate in a set of directions, with the chance of having a beam that hits us being not very great; perhaps one-tenth. Therefore, you see only 1/10 of the pulsars in range, and it is then unlikely for you to see a pulsar with more than one beam. I think that is about right since of the forty odd pulsars only four have multiple beams."

S.F. SINGER: "You mentioned two ancient problems in space physics. One is the problem of the dispersal of a bunch of particles bouncing between mirror points on a line of force. I recall that what happens is that the phase space volume gets inverted upon reflection and the bunch spreads out."

GOLD: "That's somewhat similar to what I'm talking about, it may be the analagous discussion."

A.M. LENCHEK: "That's what happens when the radiation is negligible. We don't know what happens when the radiation reaction is significant."

H. OGLEMAN: "It seems to me when you try to explain something like the spin up of the Vela pulsar you lose the advantage of a compact, rigidly rotating object if you put the emission process near the critical radius."

GOLD: "No, no. If I had a non-rigid body different parts would rotate at different velocities and if there's a jump in the rotation speed it wouldn't be the whole star that would do it, it would be some layer and the other part not. The field is largely straight in the region between the star and the critical radius so when the star jumps in frequency the pulses may show a phase jitter, but their repetition frequency has to follow the rotation of the star precisely."

REFERENCES

Drake, F.D., and Craft, H.D. 1968, Nature, 220, 231.
Gold, T. 1968, Nature, 218, 731.
Gold, T. 1969, Nature, 221, 25.
Goldreich, P., and Julian, W.H. 1969, Ap. J., 157, 869.
Ostriker, J.P., and Gunn, J.E. 1969, Ap. J., 157, 1395.
Pacini, F. 1968, Nature, 219, 145.
Ruderman, M. 1968, Nature, 218, 1128.
Salpeter, E.E., and Hamada, T. 1961, Ap. J., 134, 683.

December 1, 1969

NEUTRON STARS

H. Y. Chiu

Institute for Space Studies
Goddard Space Flight Center, NASA
New York, New York

The existence of neutron stars was first postulated by Landau (1932). The theoretical structure of neutron stars was later studied in detail by Oppenheimer and Volkoff (1939). Since then, neutron stars have been studied mainly as a theoretical object. A complete survey of literature is not intended for an article of this size. A review article by Cameron (1970) covers much of the literature on this subject.

Current theories of stellar evolution depict the life of a star as follows: In main sequence stars hydrogen reactions (pp sequence of C-N cycle) convert hydrogen into helium at a temperature of 10^7 °K, releasing energy at a rate of 25 MeV per helium nucleus formed. After a sizable fraction of hydrogen (ranging from 10% to 50% of the total mass of the star) is converted into helium, the star changes its structure, resulting in a highly condensed helium core (density $\sim 10^6$ g/cm^3), a hydrogen burning shell, an extensive hydrogen envelope whose radius is of the order of 10^{13} cm. Such stars are called red giants. As the mass of the helium core increases, the temperature increases, and eventually when it reaches 10^8 °K, helium reaction takes place, converting three helium nuclei into a carbon nucleus. Subsequent reactions are not known in detail, but carbon itself again becomes a potential nuclear fuel at a temperature of 6×10^8 °K, and at a temperature of 4×10^9 °K all elements reach the equilibrium state, the main composition of which is iron if the temperature is less than 7×10^9 °K.

Above a temperature of 6×10^8 °K, neutrino emission processes via one of the three following processes become important (Chiu, 1965)

plasma process $\gamma \rightarrow \nu + \bar{\nu}$

photoneutrino process $e^- + \gamma \rightarrow e^- + \nu + \bar{\nu}$

electron annihilation process $\gamma \leftrightarrow e^- + e^+ \rightarrow \nu + \bar{\nu}$

Above a temperature of 4×10^9 °K, the neutrino rate (in particular, that due to the electron annihilation process) is so large that the thermal energy of the star can be dissipated within a matter of seconds. Also, above a temperature of about 7×10^9 °K, the equilibrium configuration shifts from iron to helium. In either case the star contracts so rapidly that hydrostatic equilibrium cannot be maintained, and stellar collapse takes place, causing matter to be compressed into extreme densities. Neutrinos can transfer energy produced in the collapsing core to the outer parts of the star and ejecting matter. If the remaining core does not exceed a certain mass, it can remain in this highly condensed state to form a neutron star.

At high densities fermions exhibit the phenomenon of degeneracy. As a result of high density, the interparticle spacing r_e decreases, and when r_e approaches the de Broglie wave length λ_D, condensation takes place in the momentum space (degeneracy). The theory of degenerate fermion gases has been very well developed (Chandrasekhar, 1939; for recent work see Chiu, 1968); hence the following discussion will stress the qualitative features of a Fermi gas.

The uncertainty principle states that the state of a particle is defined to within uncertainties, in coordinate $\Delta\underline{r}$, and in momentum $\Delta\underline{p}$, such that

$$\Delta x\, \Delta p_x \geq \hbar \text{ , etc.} \tag{1}$$

The number densities of a fermion gas n (which follows the Pauli exclusion principle) is roughly $(\Delta x)^{-3}$; hence the relation between n and the average momentum of a particle p in the extreme degenerate case ($p \sim \Delta p$) is:

$$n \approx (p/\hbar)^3 \tag{2}$$

The kinetic energy per particle ε is:

$$\varepsilon(p) \simeq \left([(p/mc)^2 + 1]^{1/2} - 1 \right) mc^2 \longrightarrow pc(p >> mc, \text{ relativistic}) \tag{3a}$$

$$\longrightarrow p^2/2m(p << mc, \text{ nonrelativistic}) \tag{3b}$$

Electrons become relativistic when $\varepsilon \gtrsim 0.5$ MeV. From (2) and (3) we find the relation between the electron density n and the average electron energy as:

$$\varepsilon = n^{1/3}\hbar c \approx 0.2(n_{30})^{1/3} \text{ MeV} \tag{4}$$

(valid for the relativistic case $\varepsilon \gtrsim 0.5$ MeV)

where $n_{30} = n/10^{30}$ [$n_{30} = 1$ corresponds to a matter density of 3×10^6 g/cm^3 if $A/Z \sim 2$ for the nucleus component].

As density increases, the average electron energy increases, and inverse beta reaction

$$e^- + (Z,A) \rightarrow (A-1,A) + \nu \tag{5}$$

takes place, causing the nucleus to become more neutron-rich. As a result of high temperature conditions, nuclear reactions will take place, reshuffling the structure of the nucleus. However, the A/Z ratio never exceeds 3 and when $n_{30} \sim 10^6$ (matter density $\sim 10^{12}$ g/cm^3), the average electron energy is 24 MeV. The nuclear binding energy per nucleus is of the order of 8 MeV; hence the electrons are energetic enough to cause the dissociation of all nuclei, and these electrons are then captured by protons to form neutrons:

$$e^- + p \rightarrow n + \nu \qquad (6)$$

A small fraction of matter remains in the electron and proton state to raise the Fermi energy of the electrons above the energy difference between the neutron and the proton so that the neutrons cannot decay.

A detailed knowledge of the equilibrium composition at various densities requires a detailed knowledge of nuclear structure off the stability line in the Z vs. A curve. A theoretical composition at several typical densities based on the von Weizäcker mass formula has been obtained by Salpeter, and his result is shown in Table 1 (Salpeter, 1961).

In order for free neutrons to exist, the density must exceed 10^{11} g/cm^3. The density of a degenerate star increases with its mass; hence there exists a lower limit for the mass of neutron stars. Because the neutron-proton ratio is very sensitive to the density near the nucleus↔neutron transition region, the lower mass limit must be obtained by numerical integration. The numerical value of this lower mass limit is in the neighborhood of 0.18 $\odot$ ($\odot$ = solar mass = 2×10^{33} g); the central density at this mass is about 10^{14} g/cm^3. The value of this limiting mass is somewhat dependent on the nuclear interaction.

There is an upper mass limit characteristic to the neutron star. This is the mass limit set up by general relativistic conditions. Because of high density, the general relativity parameter GM/Rc^2 (where M is the stellar mass, R the radius, G and c are gravitational constant and velocity of light, respectively) is close to 0.1, and general relativistic effects must be taken into account. The most important consequence for stellar structure is: [ρ = density, $M(r)$ = mass enclosed within a sphere of symmetry of radius r].

$$\frac{dP}{dr} = -G\,\frac{(\rho + P/c^2)(M(r) + 4\pi r^3\, P/c^2)}{r^2(1 - 2GM(r)/rc^2)} \qquad (7)$$

This may be compared to the corresponding nonrelativistic case:

$$\frac{dP}{dr} = -G\,\frac{\rho M(r)}{r^2} \qquad (8)$$

We see that the effect of general relativity is to require a greater pressure gradient to achieve equilibrium. However, a greater pressure gradient requires a greater pressure, which in turn requires an even stronger pressure gradient. Therefore, there exists an upper mass limit beyond which the equilibrium configuration becomes unstable. This is characterized by the fact that in the stellar mass vs. central density curve a maximum occurs, as shown in Figure 1 (Tsuruta and Cameron, 1966). However, the value of the upper mass limit is very sensitive to the effect of nuclear interaction on the equation of state. Without correction

TABLE 1

THEORETICAL COMPOSITION OF NEUTRON STAR MATTER (after Salpeter, 1961)

(Z, A)	(26,56)	(28,62)	(28,64)	(28,66)	(28,68)	(30,76)	(30,78)	(30,80)	(32,90)	(38,120)
ϵ_F(MeV)	0.6	2.5	3.9	6.1	7.0	8.5	9.5	14.8	20.6	24
$\log\rho$	7.15	8.63	9.15	9.69	9.87	10.13	10.28	10.84	11.28	11.53

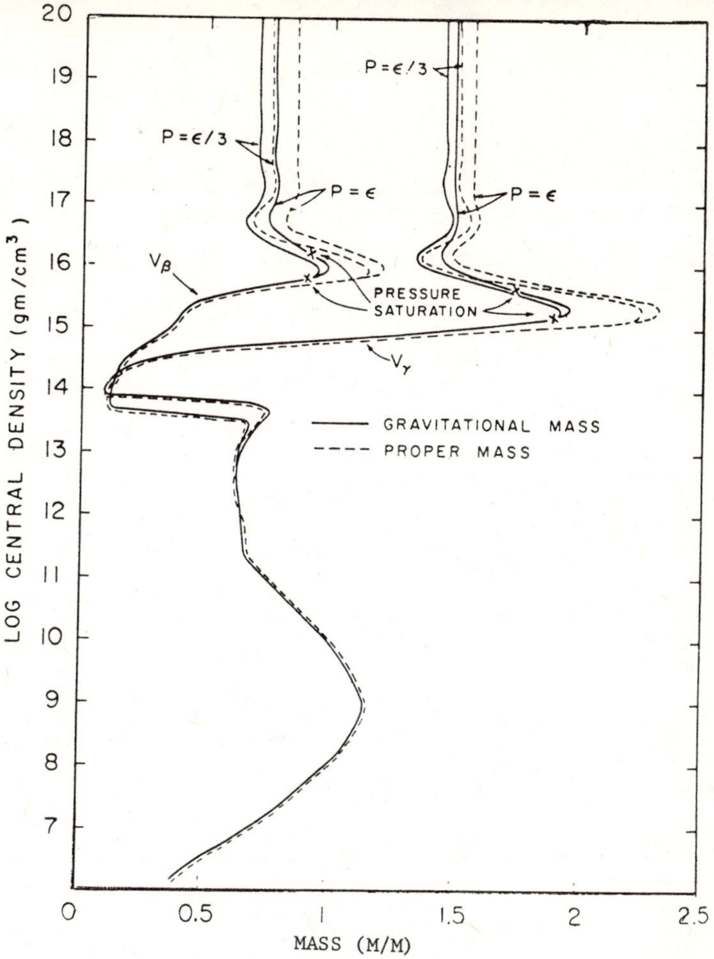

Fig. 1. The gravitational and proper masses of the models of type V_β and V_ν with the asymptotic equation of state of either $P = \varepsilon$ or $P = \varepsilon/3$. The points at which the composite equations of state switch over to the asymptotic equations of state are shown by crosses. (Tsuruta and Cameron 1966.)

for nuclear interaction (i.e., a perfect Fermi gas), the upper mass limit is 0.76$\odot$, but with nuclear interaction the upper mass limit may be as high as 3$\odot$. Unfortunately, the variations in the value of the upper mass limit corresponding to experimental uncertainties of the nuclear interaction are still quite large (1$\odot$ to 3$\odot$). Further, when the density of the neutron gas exceeds 10^{16} g/cm^3, the threshold for the production of strange particles is reached, and virtually nothing can be said about such an equation of state. At present the upper mass limit is believed to lie somewhere between 1 and 3$\odot$.

The radius of most neutron stars is of the order of 10 Km with $GM/Rc^2 \sim 0.1$.

The density of a stable star must decrease outwards, and at densities below 10^{12} g/cm^3 the composition is a mixture of neutron-rich elements. The typical structure of a neutron star with a radius of 10 Km is as follows: From center to 9 Km, neutrons (with approximately 1-6% of protons and electrons), from 9 Km to 10 Km minus 1 m, neutron-rich elements from ${}_{38}Y^{120}$ (at 9 Km where the density is $\sim 10^{12}$ g/cm^3) to Mn^{57} (at 10 Km-1 m where the density is $\sim 10^7$ g/cm^3), and most likely Fe^{56} (from 10 Km-1 m to 10 Km). The scale height (the distance over which the density changes by a factor of 2) is of the order of 1 cm at the surface.

There are several important properties of neutron matter which are discussed below:

(a) Superfluid state. In the core of a neutron star where the densities are more than 2×10^{14} g/cm^3, but below the normal nuclear density (1.2×10^{15} g/cm^3), neutrons and protons behave as a compressible fluid. The force between those neutrons lying at the top of the Fermi sea is attractive (attractive potential), and pairs of neutrons can form a bound state. Such bound neutron pairs behave like bosons, and can exhibit superfluid phenomena. Because of this bound state, in the single particle energy spectrum an energy gap of the order of 1 MeV exists. This energy gap will decrease the heat capacity of the neutron gas (if the neutron gas were an ideal Fermi gas, the heat capacity would be proportional to T). This superfluidity arises from the ^1S_o (S singlet state) of the nuclear interaction, which is attractive at distances larger than the nuclear separation in normal nucleus ($\sim 1.2\times10^{-13}$ cm) but is repulsive at smaller distances. However, other nuclear interactions may be attractive, so that the superfluidity state may persist even at higher densities.

(b) Crystalization of matter at subnuclear densities. In a neutral plasma the ions and electrons are randomly distributed, but as the temperature decreases, the Coulomb interaction eventually becomes dominant. The state of the lowest energy is one in which the ions form lattices, and the lattice structure that gives rise to the lowest Coulomb energy is the body-center lattice. At a density of 10^{14} g/cm^3 the Coulomb energy amounts to about 1 MeV per ion. It is then expected that when $kT < 1$ MeV, crystalline structure is expected. The melting temperature of such a solid has been estimated to be around 10^9 °K.

The outer layers of a neutron star are therefore expected to be composed of crystalline solids in the form of crusts. Such a crust is capable of supporting a certain amount of strain. Ruderman (1968) showed that as the rotation of a neutron star slows down, the gravitational equilibrium configuration changes slightly, introducing strain in the crust region. When the strain exceeds the amount that can be supported by the crust, sudden readjustment or "starquakes" will take place, and the rotation period of the star will change suddenly. Ruderman has interpreted the sudden decrease of the period of the pulsar Vela (and others) as resulting from such a starquake.

(c) Rotation. During the formation of a neutron star, the progenitor star contracts by a factor of 10^5 (from 10^{11} cm to 10^6 cm). If angular momentum is conserved, the rotation period P is proportional to R^{-2}, where R is the radius. The rotation period of a star like our sun is 26 days $\approx 2.2\times10^5$ seconds. After collapse into a neutron star, the sun would have a theoretical period of 10^{-5} seconds. However, the rotation period of a star is limited by the condition that at the equator the centrifugal force v_{eq}^2/R must not exceed the gravitational force GM/R^2, where v_{eq} is the velocity of rotation at the equator; otherwise matter will be ejected into space, thus decreasing the rotational period. For neutron stars, since $v_{eq} = 2\pi R/P$, we find

$$P > \sqrt{2\pi}\,(R^3/GM)^{1/2} \approx 3\times10^{-4}\,\frac{(R/10\text{Km})^{3/2}}{(M/\odot)^{1/2}}\ \text{seconds} \qquad (9)$$

Therefore, it can be concluded that even for a slow rotating star like our sun, after collapse the theoretical period is considerably shorter than that allowed by stability criteria. To dissipate the extra angular momentum, either ejection of matter or even possibly fission into two neutron stars is possible. It is also reasonable to assume that the rotational period of freshly formed neutron stars is given by equation (9).

(d) Magnetic field. The average magnetic field of many stars is in excess of 500 gauss, which is the present limit of detectability. However, the strength of the field may vary greatly over the surface. For example, although the average magnetic field of our sun is only 1 gauss, in various regions (sunspots) the field strength exceeds a few thousand gauss. During collapse, if magnetic fluxes are conserved, the field strength B also increases as R^{-2}. Assuming an average field of 100 gauss in the progenitor star, the field in a neutron star may be as high as 10^{12} gauss.

Inside such a strong magnetic field the properties of matter are different, largely due to the quantization of electron orbits. This has been studied by Chiu and Canuto in a series of papers.

The first indirect observational support for such a field is in the case of the Crab Nebula pulsar. The total energy output rate from the Crab Nebula is 10^{38} ergs/sec, and this can be accounted for by the rate of slowing down of the pulsar in the nebula. However, in order to provide a link between the loss of rotation energy and the supply of energy into the nebula, a magnetic field is needed (Ostriker and Gunn, Pacini) and the strength of the field required is about 5×10^{12} gauss.

Just after the formation of the neutron star, the internal temperature may be as high as 10^{10} °K, but cooling by neutrino emission at this stage is very rapid. Roughly after one year the internal temperature of the star falls below 5×10^{8} °K.

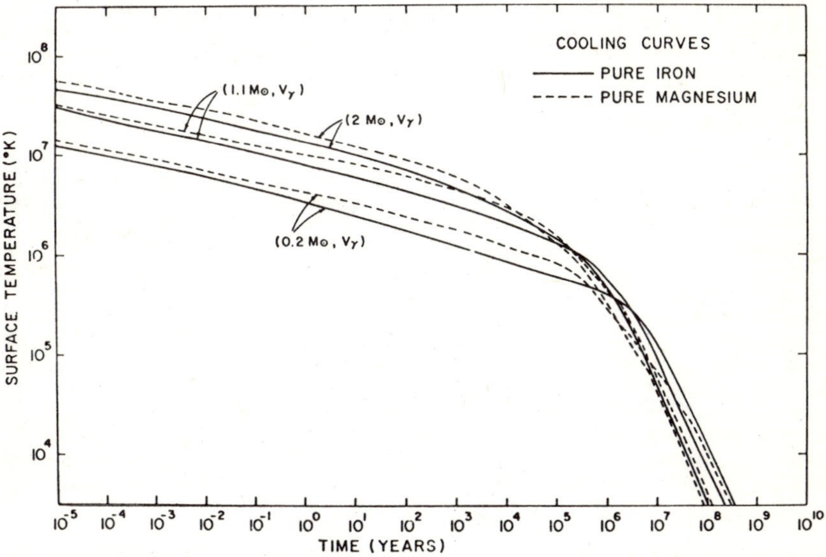

Fig. 2. The effect of composition on cooling curves. Surface temperature of models of the V_ν type potential and with different mass values are plotted as functions of time in years for iron atmospheres (solid curves) and magnesium atmospheres (dashed curves). (Tsuruta and Cameron 1966).

Nonmagnetic and nonrotating neutron star models have been considered by Chiu and Salpeter, Tsuruta and Cameron, and others. The best cooling rate has been that computed by Tsuruta and Cameron, and their results are shown in Figure 2.

The main source of opacity considered in these models is the electron scattering. The source of energy is the residual thermal energy of a neutron gas treated as an ideal Fermi gas. However, Canuto showed recently that the electron scattering opacity in a magnetic field is very anisotropic and along the direction of the magnetic field the opacity is nearly zero. Further, superfluidity in the core and crystalization in the crust may strongly reduce the heat capacity. Therefore, the cooling rates of a realistic neutron star must be regarded as unknown.

REFERENCES

Cameron, A.G.W. 1970, "Neutron Stars," to be published in Annual Review of Astronomy and Astrophysics.

Chandrasekhar, S. 1939, An Introduction to the Study of Stellar Structure (Chicago: U. of Chicago Press).

Chiu, H.Y. 1965, "Neutrinos in Astrophysics and Cosmology," Annual Review of Nuclear Science, Vol. 15.

Chiu, H.Y. 1968, Stellar Physics (Waltham, Mass.: Blaisdell Publishing Co.)

Landau, L. 1932, Phys. Zeits. Sowjetunion, 1, 285.

Oppenheimer, J., and Volkoff, G.M. 1939, Phys. Rev., 55, 374.

Ruderman, J. 1968, Nature, 223, 597.

Salpeter, E.E. 1961, Ap. J., 134, 669.

Tsuruta, S., and Cameron, A.G.W. 1966, Nature, 211, 356.

November 17, 1969

Revised October 1, 1970

NEUTRON STARS, PULSAR RADIATION
AND SUPERNOVA REMNANTS

F. Pacini*

Center for Radiophysics and Space Research
Cornell University

ABSTRACT

We summarize here a pulsar model based upon the rotation of a neutron star with a skew magnetic field.

Close to the stellar surface, an electric field of order $E \sim (v/c) B$ causes an outflow of charged particles from the star. These particles are somewhat accelerated in the near magnetosphere and become relativistic ($\gamma = E/mc^2 >> 1$) close to the speed of light distance $r = c/\Omega$. The pulsed radio emission is connected with an approximate corotational motion and is peaked at the frequency $\Omega\, \gamma^3$. In the case of NP 0531, the observed intensity of emission requires a density of order 10^{11} cm^{-3}, a Lorentz factor $\gamma \sim 200$ and a correlation length of about 10 cm.

The optical and the X-ray emission from NP 0531 can be explained as a concomitant synchrotron radiation by the same particles in a field $B \sim 10^6$ gauss. The low frequency cut-off at about 10^{15} Hz is predicted by the theory of synchrotron reabsorption and can be taken as a proof for an emission close to the speed of light distance.

The same particles responsible for the pulsar radiation can be accelerated further at $r > c/\Omega$ and they become responsible for an eventual high energy activity in the Remnant. Some evolutionary considerations about this activity are finally developed.

1. Introduction

For many years, some of the outstanding problems in astrophysics have been connected with the release of large amounts of nonthermal energy from very limited regions of space. A striking example of this type of phenomena in our galaxy is the Crab Nebula, the diffuse Remnant of a stellar explosion which occurred about 900 years ago. Its radio emission, its optical light and, most likely, its X-rays originate from the motion of relativistic electrons (energies up to 10^{13} eV, perhaps higher) in a general magnetic field $10^{-3} - 10^{-4}$ gauss. For most of these electrons, the lifetime against radiation losses is shorter than the age

*On leave from Laboratorio di Astrofisica, Frascati (Italy)

of the Nebula and they cannot have been produced only in the initial event. There should be instead a continuous supply of relativistic particles, at a rate roughly matching the total nebular luminosity (about 10^{38} ergs sec^{-1}). What is the source of this activity in the Crab Nebula?

Although the details of the Supernova phenomenon are controversial, it is widely agreed that, at least in some cases, the explosion should leave behind a condensed stellar core, the neutron star, and an expanding gaseous Remnant. A typical neutron star has a mass $M \sim 1 M_\odot$, a radius $a \sim 10^6$ cm, an average density $\rho \sim 5 \times 10^{14}$ g cm^{-3}, a gravitational binding energy about 10^{53} ergs. Also, the strength of the stellar magnetic field greatly increases during the collapse of the central core because of flux conservation: the final value can be as high as 10^{10}-10^{15} gauss, depending upon the initial conditions (Ginzburg, 1964; Woltjer, 1964).

A newly born neutron star should be expected to be a very excited object since its initial temperature can exceed 10^9 - 10^{10} °K and it can rotate or vibrate at frequencies $\Omega_{rot} \lesssim \Omega_{vib} \sim 10^3$-$10^4$ sec^{-1}. The energy content in these mechanical modes can be almost as high as the gravitational energy and is therefore comparable to the original nuclear resources of the pre-supernova stellar core. A new life begins for the star after the collapse: the first life was largely charterized by thermal phenomena, the second will be essentially nonthermal.

Heat content and vibrations disappear fast since they are subject to neutrino losses; furthermore, any small amount of rotation will mix the radial and the non-radial pulsations, thus resulting in the emission of gravitational waves. Early attempts to connect the post-outburst activity in Supernova Remnants with the presence of a hot, vibrating, neutron star were therefore unsuccessful.

Since the magnetic energy is likely to be (relatively speaking) almost negligible, one is left with only two possible stockpiles of energy available for a long time after the explosion: the neutron star rotation or the energy stored in the ions ejected in the early phases. Both possibilities were briefly mentioned as deserving investigation in a review article by Wheeler (1966) but no mechanism was suggested for the use of either of them. On the other hand, a young neutron star is likely to be a system where the magnetic and the rotation axis do not coincide, since it results from a catastrophic, probably disordered, phenomenon. This implies the possibility that the star will emit large amounts of magnetic dipole radiation at the basic rotation frequency Ω. With the magnetic fields and angular frequencies expected in a neutron star, magnetic dipole radiation can be very important and represent an appropriate coupling between the stellar rotation and the circumstellar plasma. It was therefore suggested that the Crab Nebula could be fed by a rotating neutron star emitting large amounts of magentic dipole radiation (Pacini, 1967).

A few months later, the discovery of pulsars brought an independent suggestion by T. Gold that these objects had to be identified with spinning neutron stars (Gold, 1968) and a new discussion of magnetic dipole radiation in relation to pulsars and to the activity of Supernova Remnants (Pacini, 1968; Gunn and Ostriker, 1969a).

As is well known, the discovery of fast pulsars in young Supernova Remnants and the realization that the Crab Nebula pulsar loses rotational energy at a rate roughly matching the steady nebular luminosity have lent strong support to the idea that both the pulsars and the high-energy activity in Supernova Remnants are connected with the presence of a rotating neutron star.

Several people have recently considered various aspects of the electrodynamics of rotating neutron stars. In particular, Gunn and Ostriker (1969a,b) have extended the analysis of magnetic dipole radiation and concluded that indeed very high energies can be reached by particles moving in this low frequency field. Goldreich and Julian (1969) (see also Michel, 1969) have investigated other aspects of the electrodynamics and it appears that very high energies can be reached also in the axisymmetric case. Since this work is discussed in this series of Lectures by the authors, we shall not describe it here in much detail. It is however worth stressing that many quantitative predictions (in particular, the rate of energy loss) are exactly the same in both models. We are convinced that the final theory will have to bring together pieces of information which are still considered in different frameworks. Because of this reason, we shall try to discuss mostly those aspects of the pulsar theory which do not rely upon the exclusive adoption of one of the existing models. In particular, after a brief discussion of magnetic dipole radiation, we shall concentrate on the nature of the pulsar radiation and on the relation between pulsars and Supernova Remnants.

2. The Electromagnetic Field Around a Rotating Neutron Star.

The existence of a periodical phenomenon connected with rotating neutron stars suggests that we are dealing with an asymmetric system. We shall assume that the asymmetry is related to the distribution of magnetic fields over the stellar surface: a typical example would be the presence of a field component perpendicular to the rotation axis. In this case, the perpendicular magnetic field can corotate with the star only in the near magnetosphere, at distances $r < c/\Omega$. Beyond the so-called "speed of light distance" $r_{cr} = c/\Omega$, the field becomes a radiation field at the basic frequency Ω. The rate of radiation is roughly given by

$$L \sim \frac{B^2}{8\pi} c 4\pi \left(\frac{c}{\Omega}\right)^2 \tag{2.1}$$

where $B = B(r_{cr})$. In the simple case of an inclined dipole, we have $B(r) = B_o\, a^3/r^3$ where B_o is the surface field; therefore the energy loss becomes

$$L \sim \frac{1}{2c^3} B_o^2\, a^6 \Omega^4 \tag{2.2}$$

A detailed analysis of the electromagnetic field around an oblique rotator confirms this intuitive result with a small modification in the numerical coefficient (Deutsch, 1955)

$$L = \frac{2}{3c^3} B_o^2\, a^6 \Omega^4 \tag{2.3}$$

We would like to emphasize that the existence of this form of energy loss does not depend upon a dipole geometry but only upon the idea that we are dealing with an asymmetric system; the relation $L \propto \Omega^4$ does however require a dipole field.

If the condition $B^2/8\pi > nmc^2$ is not verified in the magnetosphere, we should expect corotation of the near field only up to the "speed of Alfven distance" i.e., up to a distance r_{cr} such that $\Omega r \lesssim \Omega r_{cr} \sim v_A$, the Alfven velocity. Considerations similar to the one made before give in this case a magnetohydrodynamic energy loss

$$L \sim \frac{B^2}{8\pi} v_A 4\pi \left(\frac{v_A}{\Omega}\right)^2 \tag{2.4}$$

(the explicit value of B at $r = r_{cr}$ depends upon the details of the magnetosphere).

In the following, we shall assume a dipole field perpendicular to the rotation axis, corotating with the star up to about the speed of light distance. At $r \sim r_{cr} = c/\Omega$ retardation effects come into play and the field lines which emanate from the magnetic poles are swept back: the resulting tension corresponds to a torque reducing the stellar rotation. If the moment of inertia of the star does not change, the star should slow down and one predicts the following period evolution (Pacini, 1968)

$$P^2 = P_o^2 + A^2 B_o^2 t \tag{2.5}$$

where P_o is the initial period and $A^2 \sim 2 \times 10^{-39}$ gauss^{-2} sec (for a standard neutron star). We can also define the quantity $T_o = P/2\dot{P}$ in terms of present observational data; one easily sees that at a time $t = -T_o$ the period was very small and therefore the age of a pulsar is given by T_o.

The evolution of the pulsar rotation can be divided into two distinct phases. For a time

$$t \lesssim T_{cr}, \quad \text{where } T_{cr} = \left(\frac{P_o}{AB_o}\right)^2,$$

the period does not change appreciably; for $t > T_{cr}$ we have $P \propto B_o t^{1/2}$. Numerically

$$T_{cr} \sim 10^7 \left(\frac{P_o}{10 \text{ msec.}}\right)^2 \left(\frac{10^{10} \text{ gauss}}{B_o}\right)^2 \text{ years.} \tag{2.6}$$

We note that for low magentic fields (say, $\sim 10^{10}$ gauss), there is enough time for the gaseous Supernova Remnant to fade away before any appreciable increase in the period of the central pulsar. The existence of very fast pulsars not associated with gaseous Remnants is therefore a definite possibility. At a lower power level, such pulsars could resemble the Crab Nebula source NP 0531 and emit mostly in the optical or X-ray band. No pulsar of this kind has yet been discovered but a search among the known X-ray sources appears desirable (Pacini, 1969b).

On the other extreme, if the magentic field B_o is much stronger than 10^{10} gauss, the phase characterized by a constant period is very short and is followed by an evolution $P(t) \propto B_o t^{1/2}$. An upper limit to the period can result if the magnetic field decays over a time scale $10^6 - 10^7$ years (Grewing and Priester, 1969; Gunn and Ostriker, 1969c; Pacini, 1968, 1969a). Some recent investigations of the conductivity in the neutron star matter tend however to suggest a much longer lifetime (Baym et al., 1969); also, the existence of a pulsar with an age in excess of 10^8 years (as deduced from the slowing down rate) has been recently reported (Hunt, 1969). It appears therefore that the field decay is important only for neutron stars having a large crystalline mantle (low mass neutron stars) and a relatively low conductivity.

A determination of the slowing down rate yields the value of the perpendicular component of the magentic field through the relation

$$B_o = \frac{1}{A}\left(2P\frac{dP}{dt}\right)^{1/2} \sim 2 \times 10^{19}\left(P\frac{dP}{dt}\right)^{1/2} \text{ gauss.} \qquad (2.7)$$

For all pulsars with measured slowing down rate, one finds $B_o \sim 10^{11} - 10^{12}$ gauss. Although some authors regard this concentration of magnetic field strengths about a definite value as real, it seems to us likely that it is due to selection effects. Clearly, if $B_o \ll 10^{12}$ gauss the slowing down is very small and it will take a long time to measure the slowing down rate; if $B_o \gg 10^{12}$ gauss, the lifetime of the pulsar is very short.

Of considerable interest for the theory of pulsars and Supernova Remnants is the expression of the electromagnetic stresses at large distances $r > c/\Omega$ from the star. Both the radial stresses P_r and the tangential stresses P_ϕ are periodic and are given by (Bertotti et al., 1969a)

$$P_r = \frac{B_o^2}{4c^4}\frac{a^6}{r^2}\Omega^4\left\{1-\sin^2\theta\cos^2\left[\Omega\left(\frac{r}{c}-t\right)+\phi\right]\right\} \qquad (2.8)$$

$$P_\phi = \frac{B_o^2}{2c^3}\frac{a^6}{r^3}\Omega^3\left\{\sin\theta\sin^2\left[\Omega\left(\frac{r}{c}-t\right)+\phi\right]\right\} \qquad (2.9)$$

θ and ϕ are the polar coordinates. The rotation pole corresponds to $\theta = 0$. The asymmetry of these stresses could be reflected in the shape of SN Remnants (at least as long as the precession of the rotation axis can be neglected).

It is very important to recall that the surface of the star and the near magentosphere are dominated by an electric field $E \sim (v/c)\cdot B$ where $v \sim \Omega r$ (Goldreich and Julian, 1969). There is, as a consequence, a continuous outflow of charged particles from the star. These particles should be somewhat accelerated in the near magnetosphere and are expected to become relativistic at $r \sim c/\Omega$. At $r > c/\Omega$, an electric field $E \sim B$ gives the possibility of accelerating particles to a maximum energy of order $E\ (c/\Omega)$ eV. For the Crab Nebula this amounts to about 10^{14} eV.

The magnetic dipole radiation theory leads to the expectation of a large scale magentic field B in the circumpulsar space. At a distance r from the star its value is given by (Pacini, 1968)

$$B(r) \sim \frac{B_o a^3}{c^2} \frac{\Omega^2}{r} \tag{2.10}$$

In the case of the Crab Nebula, the resulting value $B \sim 10^{-4}$ gauss is in good agreement with the estimated strength. (We thank Dr. Gunn for pointing out this agreement to us).

Finally, we stress again that Eq. (2.5), (2.6), (2.7) and (2.10) which have been considered here in the framework of the magnetic dipole radiation theory hold also in the axisymmetric case. The underlying physics is different and is discussed in the Lecture by Dr. Goldreich.

3. The Nature of Pulsar Radiation.

The amount of energy lost by a neutron star as pulses of radiation is a relatively small fraction of the total loss of rotational energy. In the case of the Crab Nebula pulsar NP 0531, taking into account the entire electromagnetic spectrum, the efficiency in producing pulses is of the order of 10^{-2}-10^{-3}; most of the energy loss goes into relativistic particles. For long period pulsars the efficiency is of the same order but, in particular cases, might perhaps be somewhat higher. One should however be careful in evaluating the efficiency from the slowing down rate; different neutron stars might have moments of inertia differing by more than a factor ten.

No detailed explanation for the relative efficiency of the pulsar emission compared to the acceleration process is available at present. However, the observed trend is consistent with the idea that the pulsed radiation process occurs close to the speed of light circle while most of the rotational energy is transferred to the particles beyond the critical distance. The pulsar radiation itself can only involve (at most) the fraction of energy transferred to the particles before (or close to) the critical distance.

Although the pulses are not significant from an energetic point of view, an understanding of the radiation mechanism would provide much information about the neutron star magnetosphere. In the following, we shall present a scheme where radiation occurs close to the speed of light distance. The basic ideas were developed in collaboration with Dr. Bertotti and Dr. Cavaliere, soon after the discovery of optical and X-ray emission from NP 0531 (Bertotti et al., 1969b,c). Some new observations have become available recently and therefore we shall add here some new remarks.

We shall consider the coherent radioemission to be the result of a semi-corotational motion of charges in the upper magentosphere of the neutron star, at a distance of order c/Ω. This is similar to what has been suggested by Gold (1968, 1969) but we stress that no strictly corotational kinematics is required. It is enough to assume that the particles, before leaving the surroundings of the speed of light distance, are already relativistic and that the radius of curvature of their orbits is of order c/Ω. On the other hand, the optical and X-ray emission from NP 0531 can be explained as incoherent synchrotron radiation. In the case of this pulsar, the required densities will turn out to be in rough agreement with the value to be expected at $r \sim c/\Omega$ if the neutron star magnetosphere is the source of the particles injected into the Crab Nebula.

The main characteristic of the radio emission from pulsars is the very high brightness temperature T_b, usually of the order of 10^{21} °K (for the Crab pulsar NP 0531, $T_b \sim 10^{25}$-10^{26} °K). This rules out any mechanism based upon particles radiating independently since the thermodynamic limitation

$$k\,T_b \lesssim [\text{energy per particle}] \quad (3.1)$$

would require individual energies in excess of 10^9 ergs. The radiation process should clearly be a coherent one, with many charges radiating in phase. One of the possible ways of realizing this coherence is by bunching together many particles within a region having one dimension L (the one along the line of sight) smaller than the wavelength of the emitted radiation. For the particular case of a quasi-spherical bunch, the radiation emitted at wavelengths $\lambda > L$ is then enhanced by a coherence factor $F \sim n\,L^3$ (n is the density of radiating units). The thermodynamic limitation becomes

$$k\,T_b \lesssim (\text{energy per bunch}) \approx (nL^3)\,(\text{energy per particle}) \quad (3.2)$$

In order to have spatial coherence, one should overcome the electrostatic repulsion among the charges; the energy necessary should be provided by the surrounding medium (particles or fields). It is therefore clear that bunches can be formed and give rise to coherent radiation only in the presence of large concentrations of energy. This is indeed the case of pulsars.

Let us now assume that at $r \sim c/\Omega$ there is a sector of the magnetosphere containing bunches of charges in the energy range appropriate to give rise to radio emission. The extent of this sector is determined by the duty cycle of the pulsar.

In order to produce radio frequencies (say, at about 150 MHz) the Lorentz factor γ of the charges should satisfy the relation

$$\Omega\,\gamma^3 \sim 10^9\ \text{sec}^{-1} \quad (3.3)$$

This emission is peaked forward and the periodic appearance of the active sector (possibly above the magnetic poles, at a distance from the star of order c/Ω) can give rise to a lighthouse effect.

The usual formulae for the power radiated from relativistic accelerated charges (see, e.g., Jackson, 1962), give an energy loss per unit time

$$p \sim \frac{e^2}{c}\,\Omega^2\,\gamma^4 \quad (3.4)$$

(most of it is radiated close to the critical frequency $\Omega\,\gamma^3$). The energy emitted per unit time from a volume V containing n charges per unit volume is therefore

$$L \sim \frac{e^2}{c}\,\Omega^2\,\gamma^4\,n\,V\,(1 + F) \quad (3.5)$$

(the factor F takes into account a possible coherent enhancement).

We shall use the previous equations in order to obtain an estimate for the pulsar parameters. Once the factor γ of the particles is obtained from Eq. (3.3), the density can be deduced by comparing Eq. (3.4) with the measured energy output. Since the detailed geometry of the source is largely unknown, we shall tentatively adopt an active volume $V \sim \ell^3$ where ℓ is the active sector of the speed of light circle. Also, we shall assume an average correlation length L of about 10 cm which gives coherent emission at $\lambda > 10$ cm. The result obtained for the density n can be checked by requiring that Eq. (3.2) be satisfied. Also, if n is known independently (in the case of NP 0531 from the synchrotron emission at high frequencies) one can use Eq. (3.2) to determine the correlation length.

In the case of long period pulsars, such as for instance CP 1919, we have $\Omega \sim 10$ sec^{-1} and therefore $\gamma \sim 500$. A duty cycle of order 1/30 corresponds to an active sector of about $\ell = 3 \times 10^8$ cm and to an emitting volume $V \sim 10^{25}$cm^3. The power output L is close to 10^{28} ergs. sec^{-1}. This gives a density

$$n \sim 10^8 \left(\frac{L}{10^{28}\ \text{ergs sec}^{-1}}\right)^{1/2} \left(\frac{V}{10^{25}\text{cm}^3}\right)^{-1/2} \left(\frac{L}{10\ \text{cm}}\right)^{-3/2} \text{cm}^{-3} \qquad (3.6)$$

The number of particles per bunch is $F \sim 10^{11}$; with an energy per particles 10^{-3}-10^{-4} ergs, it permits brightness temperatures as high as 10^{23} °K.

We can now turn our attention to the electromagnetic spectrum of the Crab Nebula pulsar NP 0531 (Figure 1). As is well known, this is the only

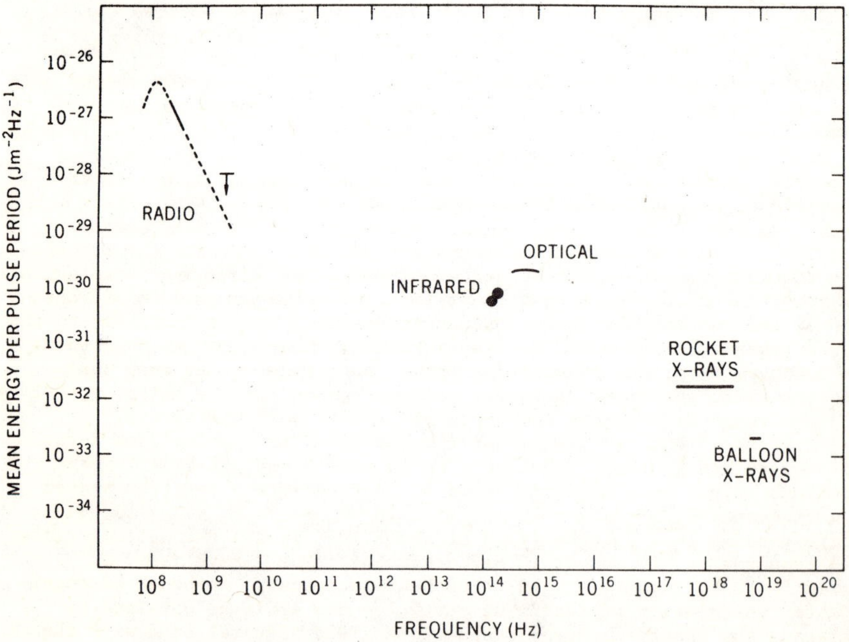

Fig. 1. The electromagnetic spectrum of the Crab Nebula pulsar NP 0531 (prepared by Dr. S. Maran).

source showing strong optical and X-ray emission. Its radio spectrum is peaked at about 100 MHz. At lower frequencies, the pulsed flux density drops off very rapidly and most of the emission comes from the compact low frequency source, also situated in the center of activity of the Crab. The spectrum of the compact source is almost an extrapolation of the high frequency portion of the pulsar spectrum. Since also the pulse width increases rapidly with decreasing frequency, it appears likely that the compact source is just a smeared image of NP 0531. The possible connection of the low frequency source with a smeared pulsar had indeed been suggested before the discovery of NP 0531 (Pacini, 1968; Woltjer, 1968). At $\nu > 100$ MHz the radio spectrum decreases again very rapidly, with a slope of order 2. Nothing is known about the emission between the high radio frequencies and the infrared but we know that the spectrum rises again toward the optical region and then decreases toward the X-rays.

The peaks of the pulses in the radio and in the high frequency part of the spectrum are simultaneous within an experimental uncertainty of a fraction of a millisecond and therefore the processes giving rise to the radio and to the high frequency emission occur in the same region. These processes are probably different but they are likely to involve the same radiating particles.

Arguments similar to the one applied to CP 1919, if transferred to the radio emission from NP 0531, give a typical corotational γ of order 200. Since most of the radiation is emitted within a few hundred microseconds, the emitting sector should be $\ell \sim 10^7$ cm. Assuming again $V \sim \ell^3 \sim 10^{21}$ cm^3 and $L \sim 10$ cm, the absolute radiopower (about 10^{31} ergs sec^{-1}) corresponds to a density

$$n \sim 10^{11} \left(\frac{L}{10^{31}\text{erg sec}^{-1}}\right)^{1/2} \left(\frac{V}{10^{21}\text{cm}^3}\right)^{-1/2} \left(\frac{L}{10 \text{ cm}}\right)^{-3/2} \text{cm}^{-3} \qquad (3.7)$$

The number of particles per bunch is roughly 10^{14}: since the energy per particle is about 10^{-4} ergs, we see that we can indeed reach $T_b \sim 10^{26}$ °K. We also note that the formal plasma frequency $\omega_p = 5.64 \times 10^4 \sqrt{N_e}$ exceeds the frequency of the emitted radiation. The reliability of the usual theory is however doubtful in our case. We are dealing with a purely relativistic magnetosphere embedded in a very strong magnetic field ($\omega_H >> \omega_p$) and the particles largely behave in a collective manner.

The observed spectral distribution probably reflects the correlation function (which gives the degree of coherence at a given wavelength) more than the energy distribution of the radiating charges. Also, that part of the variability in the pulsar intensity which is intrinsic to the source probably corresponds to fluctuations in the degree of coherence. The incoherent optical emission is indeed much more stable than the coherent radio emission (the other obvious reason for the better stability at high frequencies is the unimportance of diffraction effects).

We shall now discuss an interpretation of the optical and X-ray emission as synchrotron radiation. As is well known, if relativistic electrons spiral in a magnetic field B, they radiate a spectrum peaked at a frequency $\nu_{max} \simeq 1.2 \times 10^6 B^2_{\perp} \gamma^2$. If we assume $B_{\perp} \simeq B \simeq 10^6$ gauss (corresponding to a surface field of about 10^{12} gauss) the optical radiation requires $\gamma \simeq 20$ (about 10 MeV electrons); the X-rays (say, at $\nu \simeq 10^{18}$ sec^{-1}) require $\gamma \simeq 10^3$ (500 MeV electrons). An order of magnitude for the density of radiating particles can again be obtained by comparing the observed energy output in a given spectral band with the theoretical expression

$$L \sim \frac{e^2}{c}\left(\frac{eB}{mc\gamma}\right)^2 \gamma^4\, n\, V \sim 1.6 \times 10^{-15}\, B^2 \gamma^2\, n\, V \qquad (3.8)$$

Assuming again $V \sim 10^{21}$ cm^3 and the Lorentz factor relevant to the spectral band under consideration, we find in the case of optical emission

$$n_e(\text{at } 10 \text{ MeV}) \sim 2 \times 10^{12} \left(\frac{L}{10^{33}\text{ergs sec}^{-1}}\right)\left(\frac{V}{10^{21}\text{cm}^3}\right)^{-1} \text{cm}^{-3} \qquad (3.9)$$

The X-ray emission corresponds to

$$n_e\ (\text{at } 500 \text{ MeV}) \sim 10^{11} \left(\frac{L}{10^{35}\text{ergs sec}^{-1}}\right)\left(\frac{V}{10^{21}\text{cm}^3}\right)^{-1} \text{cm}^{-3} \qquad (3.10)$$

The low frequency cut-off observed between the optical and the infrared frequencies should also be explained. A cut-off would occur if there is a deficiency of low energy electrons and indeed the monoenergetic case considered earlier (Bertotti et al., 1969b,c) leads to a behavior of this sort. A similar situation, however, appears artificial and furthermore one predicts a flux density rising at low frequencies like $\nu^{1/3}$ while the observed spectrum is certainly steeper. A much more likely explanation for the cut-off is synchrotron reabsorption. If the brightness temperature of the source becomes, at a certain frequency, comparable to the equivalent temperature of the relativistic electrons, self-absorption occurs. One has then the following relation between the angular size of the source (rad), the magentic field strength $B_\perp$, and the quantities F_{max} (in flux units) and ν_{max} which refer to the peak in the observed spectrum

$$\theta \simeq 4 \times 10^3\, F_{\nu_{max}}^{1/2}\, B_\perp^{1/4}\, \nu_{max}^{-5/4} \qquad (3.11)$$

If

$$\theta = \frac{10^7 \text{ cm}}{5 \text{x} 10^{21}} \simeq 2 \times 10^{-15} \text{ rad.}, \quad B_\perp \simeq 10^6 \text{ gauss}, \quad F_{max} \simeq 3 \times 10^{-2}$$

(during the pulse), one predicts a cut-off at $\nu \simeq 10^{15}$ Hz.

The good agreement with the observations strengthens the idea that we are dealing with synchroton radiation close to the speed of light distance.*

To summarize, a density $n_e \sim 10^{11}$-10^{12} electrons cm^{-3} close to the speed of light circle and an average energy 100 MeV can be inferred both from the theory of radio emission and from the optical and X-ray emission. It is remarkable that this density corresponds to a flux through the speed of light distance or order

$$4\pi \left(\frac{c}{\Omega}\right)^2 c\, n_e \sim 10^{40} \text{ particles sec}^{-1} \tag{3.12}$$

This value agrees with a "pre-pulsars" estimate by Shklovskii (1966) of the number of particles continuously injected into the Crab Nebula. Clearly, these particles should be accelerated at $r > c/\Omega$ to reach the energies they are known to have in the Remnant. We also stress that the success of the synchrotron interpretation of optical and X-ray emission as synchrotron radiation close to the critical distance, plus the simultaneity of this emission with the radio emission, forces us to consider also the radio emission as a process occurring at $r \sim c/\Omega$. As we have described, approximate corotation itself can possibly be the source of radio emission.

Finally, we notice that, if the typical energy of the particles at the speed of light circle is only a few hundred MeV, only very fast pulsars are likely to emit optical and X-ray radiation. In the case of long period pulsars, the speed of light circle is too far from the star and the local magnetic field is too small. An emission spectrum with two characteristic peaks (one at radio frequencies, possibly due to corotation; the other at higher frequencies, due to synchrotron process) is in our opinion a likely feature of most sources. The following evolutionary trend should however be expected: the high frequency peak (proportional to B) should gradually shift from the optical and X-ray frequencies toward the infrared; the power emitted by this mechanism (proportional to B^2) should also become progressively smaller.

4. Pulsars and Supernova Remnants.

In this section we would like to discuss some aspects of the relationship between pulsars and the high energy activity in Supernova Remnants. This discussion clearly stems from the evidence that the source of activity for the entire Crab Nebula should be identified with a rotating neutron star.

---

*In a recent independent attempt to pursue a similar interpretation of the high frequency emission as synchroton radiation, Shklovskii (1969) has used. Eq. (3.11) and ν_{max} as quoted by Oke (1969) to determine exactly $B_\perp$. He finds $B_\perp \sim 3 \times 10^3$ gauss (i.e., a small pitch angle) and therefore he must invoke a higher density to explain the observed fluxes. We would like to stress the strong dependence of $B_\perp$ on θ and ν_{max}, and the uncertainties in the correct shape of the spectrum in the optical region. In our opinion, the use of Eq. (3.11) to determine $B_\perp$ is premature; the eventuality of a small pitch angle for the particles radiating at $r \sim c/\Omega$ should however be kept in mind, especially in view of the strong radiation losses.

Understanding of Supernovae is far from being complete and the question of the circumstances under which a neutron star can be formed is controversial. Also, the known Supernova Remnants are widely different: some of them emit X-rays and have an intense non-thermal activity, some appear able to produce only radio emission; some have a diffuse structure, like the Crab Nebula, some others have a shell structure, like Cas A. We would like to sketch here a picture for the evolution of SN Remnants assuming that they contain a rotating superdense star (the corresponding pulsar might not be observable because of geometrical or other effects).

We shall first show how the global parameters of a given Nebula (luminosity and age, or luminosity and luminosity change) determine the parameters for the central star (Cavaliere and Pacini, 1970). Also, we shall show that the rough distinction among Remnants showing a diffuse structure versus Remnants with a shell structure can be accounted for by the existence of a central neutron star at a different evolutionary phase. (For a given age of the Remnant, this clearly corresponds to different magnetic field strengths on the surface of the star).

If the loss of rotational energy corresponds to the production of relativistic particles and to the steady luminosity L of the Remnant, we have

$$L \sim \frac{2}{3c^3} B_o^2 a^6 \Omega^4 \tag{4.1}$$

The luminosity L should decrease about four times faster than the rotation frequency of the neutron star

$$\frac{\dot{L}}{L} = 4 \frac{\dot{\Omega}}{\Omega} \tag{4.2}$$

The previous equations determine, as a function of L and $\dot{L}/L$, both the magnetic field B_o and the rotation frequency Ω of the central object (I is the moment of inertia)

$$B_o = \frac{I}{4\sqrt{AL}} \frac{\dot{L}}{L} \tag{4.3}$$

$$\Omega = 2\sqrt{\frac{L}{I\dot{L}}} \tag{4.4}$$

If we assume that the rotating star has an age $T_o > T_{cr}$, we can write (as discussed in Section 2)

$$T_o = \frac{1}{2} \frac{\Omega}{\dot{\Omega}} = 2 \frac{L}{\dot{L}} \tag{4.5}$$

In this case one can therefore use the age of a Supernova Remnant instead of the luminosity change to obtain the parameters B_o and Ω.

One can easily see that the parameters of the neutron star inside the Crab could have been inferred in the past using the total luminosity $L \sim 10^{38}$ ergs sec^{-1} and the age $T_o \sim 900$ years. The same method can therefore be applied with some confidence in the case of other SN Remnants if there is a suspicion of high energy activity. If this is done in the particular case of Cas A, assuming $L = \alpha\, 5 \times 10^{36}$ ergs sec^{-1} and a luminosity change $\dot{L}/L$ of order 1% year^{-1} (which also agrees with the independently estimated age of about 250 years) one obtains for a possible pulsar

$$P = \frac{0.7}{\sqrt{\alpha}} \text{ sec} \qquad B_o = \frac{2 \times 10^{14}}{\sqrt{\alpha}} \text{ gauss} \tag{4.6}$$

Even if $1 \leq \alpha \leq 10$, one puts with some confidence the limits $0.2 \lesssim P \lesssim 0.7$ sec.

Parameters showing again the presence of a relatively long period pulsar (say, 0.1 up to about 1 sec) can similarly be obtained in other historical Remnants such as the Tycho and the Kepler Supernovae. The relatively long period does not contradict the known ages (a few hundred years) because of the very strong predicted magnetic field and concomitant rapid slowing down. Among these sources, Cas A is probably the best candidate for finding a pulsar because it is relatively close. Unfortunately, the steady emission from this object (which shows a puzzling enhancement toward the center of expansion!) is very strong and therefore many hours of integration are necessary for a possibly successful search. We note that no appreciable pulsed optical or X-ray radiation should come from the Remnants Cas A, Tycho and Kepler, if indeed the eventual pulsar periods are quite long.

It is remarkable that these SN Remnants all show a shell structure rather than a diffuse structure similar to the one of the Crab Nebula. Some people have been led to argue on this basis that there is no evidence for continuous activity in them and therefore little hope of finding a pulsar. We would like to present here an argument which connects the shell structure with the very presence of a long period pulsar. According to Eq. (2.10), a pulsar with a long period can only generate a very weak magentic field over most of the volume enveloped by the shell. The predicted nebular field is always of the order of or smaller than the general interstellar field. Even if relativistic particles are generated by the central pulsar, they will not radiate very strongly over most of the volume occupied by the Remnant. Most of the emission will come from the outer shell where the magentic field is unlikely to be directly connected with the pulsar and should be expected to be stronger.

If the above point of view is correct and the presence of a shell structure (rather than a diffuse structure) indicates the presence of a long period pulsar, then one should conclude that the most important agent in determining the morphology and the evolution of a Supernova Remnant is indeed the magnetic field strength on the surface of the central neutron star. It is clear in this framework that only relatively young objects among SN Remnants with a shell structure can be expected to have an intense high energy activity and emit X-rays. When $P \gtrsim 1$ sec., i.e., after a few hundred or thousand years (depending upon the stellar B_o), the input of high energy particles given by Eq. (4.1) has become too low.

Also, the apparently "exceptional" character of the Crab Nebula appears to be just a consequence of the fact that, with a surface field of order 10^{12} gauss, the central star can remain a long lasting producer of a strong nebular field (10^{-4} gauss) and also accelerate relativistic particles up to high energies.

Finally, I want to comment on the small acceleration that was observed in the Crab pulsar in October, the "Crab quake." This was about 100 times smaller than the acceleration of the Vela pulsar and represents a change in radius of the neutron star of 10μ. It might be interesting to see if there is any connection between the occurence of such Crab quakes and the occurence of the wisps in the Crab nebula that have been seen for a long time.

5. Conclusions.

To summarize, we would like to review the most important features of the previous discussion. At the basis of our model there is a rotating, magnetized (10^{10}-10^{14} gauss) neutron star. The magnetic axis is inclined with respect to the rotation axis and therefore at $r \sim c/\Omega$ the lines of force emanating from the magentic poles are swept back. The resulting tension corresponds to a torque slowing down the stellar rotation. Close to the stellar surface an induced electric field of order $E \sim (v/c) B$ causes an outflow of charged particles from the star. These particles are somewhat accelerated in the near magnetosphere and become relativistic at $r \sim c/\Omega$. They generate the pulsating component of the radiation close to the speed of light distance, possibly above the magnetic poles. One possible mechanism for the radio spectrum is the emission from bunches of particles with a Lorentz factor of order 10^2, moving along orbits with a radius of curvature of order c/Ω. We find that the process requires a density of order 10^8 cm^{-3} in the case of long period pulsars and about 10^{11} cm^{-3} for NP 0531. The size of a bunch is a few centimeters, possibly about 10 cm.

The emission from NP 0531 in the optical and X-ray range can be easily interpreted as synchrotron radiation from electrons having an energy 10 MeV up to about 500 MeV, moving in a magnetic field $B \sim 10^6$ gauss. Again, the density turns out to be 10^{11}-10^{12} particles cm^{-3}, in agreement with the independent estimate from the radio emission. The cut-off from the optical toward the infrared emission can be explained as synchrotron reabsorption. Slow pulsars have the critical distance far from the star and therefore the magnetic field strength at $r \sim c/\Omega$ is small. The peak of the synchrotron emission should be expected to fall in the infrared but at a low power level.

The same particles which give rise to the pulsed radiation subsequently diffuse far beyond the speed of light distance and acquire more energy by moving in an electric field $E \sim B$. A rough order of magnitude for the work done by the electric field can be estimated by the product $E \times (c/\Omega)$; in the case of the Crab Nebula this leads to a maximum energy 10^{14} eV. A detailed discussion of the acceleration process in the framework of the magnetic dipole radiation theory can be found in the paper by Ostriker and Gunn (1969b). Similar performances can however be reached in the framework of the axisymmetric theory (Goldreich and Julian 1969).

Also, it is worth recalling that: (a) some neutron stars might have a magnetic field that is relatively weak (say, 10^{10} gauss). In this case the star remains with its initial fast rotation for a very long time, while the SN Remnant disappears. Some of these old, fast spinning stars could be connected with X-ray sources; (b) if the surface field is much stronger than 10^{12} gauss, the star slows down to about 1 sec before the disappearance of the Remnant. If a Remnant contains a pulsar of this sort, it should show a rapidly expanding shell structure; the nonthermal activity should be limited to the early life.

Finally, the evidence for a mechanism which very efficiently converts rotational energy into energy of relativistic particles around a neutron star has led to the suggestion that the same process could be effective around a rotating supermassive star (Morrison, 1969; Pacini, 1969; Cavaliere et al., 1969; Woltjer, 1969). The basic difference between the two cases is that a neutron star is an evolved, stable object while the supermassive star might be far from its final state. It is conceivable that the electromagnetic loss of angular momentum controls the late evolution of a supermassive object. The importance of this line of investigations for a possible understanding of the violent events in the nuclei of extragalactic systems is obvious.

DISCUSSION

J.A. EARL: "I calculate the lifetime of an electron in a million gauss field and it comes out to be about 10^{-6} sec."

F. PACINI: "The basic point here is that there is a continuous loss by radiation, as you say, but one should also expect a simultaneous acceleration by the electromagentic field which transmits to the particles the rotational energy of the star."

H. ZAPOLSKY: "I estimate that when these electrons strike the plasma, they generate about 10^{26} bremsstrahlung X-rays per second. Is this observable?"

PACINI: "I don't know. But, since the motion of most particles is likely to be coherent and directional, I am not sure if you can calculate the number of collisions as if it were isotropic."

ACKNOWLEDGMENTS

It is a pleasure to acknowledge many fruitful discussions with Drs. F. Drake, T. Gold, E. Salpeter, G. Setti and L. Woltjer. A special mention should be made of the extended collaboration with Drs. Bertotti and Cavaliere; many of the ideas expressed here are the result of this collaboration.

This work was supported partly by the National Science Foundation under Grant GP-9621 and partly by the Office of Naval Research under Contract N 00014-67-A-0077-0007.

REFERENCES

Baym, G., Pethick, C. and Pines, D. 1969, Nature, 224, 673.
Bertotti, B., Cavaliere, A. and Pacini, F. 1969a, Nature, 221, 624.
Bertotti, B., Cavaliere, A. and Pacini, F. 1969b, paper given at the I.A.U. Symposium on X- and Gamma Ray Astronomy (Rome, May 1969), to be published.
Bertotti, B., Cavaliere, A. and Pacini, F. 1969c, Nature, 223, 1351.
Cavaliere, A., Pacini, F., Setti, G. 1969, Ap. Lett., 4, 103.
Cavaliere, A. and Pacini, F. 1970, Ap. J. (Letters) in press.
Deutsch, A. 1955, Annales d'Astrophysique, 18, 1.

Ginzburg, V. 1964, Dokl. Akad. Nauk URSS, 156, 43.
Gold, T. 1968, Nature, 218, 731; 1969, Nature, 221, 25.
Goldreich, P. and Julian, W. 1969, Ap. J., 157, 869.
Grewing, M. and Priester 1969, Forschung Ber. Ast. Inst. Bonn, 69-04.
Gunn, J. and Ostriker, J. 1969a, Nature, 221, 454.
Gunn, J. and Ostriker, J. 1969b, Ap. J., 157, 1395.
Gunn, J. and Ostriker, J. 1969c, Nature, 223, 813.
Hunt, G. 1969, Nature, 224, 1005.
Jackson, J.D. 1962, Classical Electrodynamics (New York: Wiley).
Michel, F.C. 1969, Ap. J., 158, 727.
Morrison, P. 1969, Ap. J. (Letters), 157, L73.
Neugebauer, G., et al. 1969, Ap. J. (Letters), 156, L115.
Oke, J.B. 1969, Ap. J. (Letters), 156, L49.
Pacini, F. 1967, Nature, 216, 567.
Pacini, F. 1968, Nature, 219, 145.
Pacini, F. 1969a, Nature, 224, 160.
Pacini, F. 1969b, submitted to Nature.
Pacini, F. 1969, invited paper given at the Nice Conference on "Fundamental Physics and Astrophysics" (Nice, April 1969) to be published.
Shklovskii, I.S. 1968, Supernovae (New York: Wiley).
Shklovskii, I.S. 1969, paper given at the meeting on "Pulsars and High Energy Activity in SN Remnants" (Rome, December 1969).
Wheeler, J.A. 1966, Ann. Rev. Astr. Ap., 4, 393.
Woltjer, L. 1964, Ap. J., 140, 1309.
Woltjer, L. 1968, Ap. J. (Letters), 152, L179.
Woltjer, L. (unpublished, 1969).

November 11, 1969

PULSARS—ROTATING MAGNETIC NEUTRON STARS*

Hong-Yee Chiu

Institute for Space Studies
Goddard Space Flight Center, NASA
New York, N. Y. 10025

ABSTRACT

In this paper we present an account of a theory of pulsar radio emission. The emission mechanism is via maser amplification process. This theory avoids the difficulty of coherent plasma emission, that the bandwidth of radiation must be less than $\lambda/2$. The high brightness radio temperature and the insensitivity of pulsar radio flux to pulsar periods can be easily accounted for.

1. Introduction

In 1968 Hewish and Miss Bell discovered the existence of pulsating celestial radio sources (Hewish et al., 1968). These radio sources emit bursts of radio emission in the 100-1000 MHz frequency region with a $\nu^{-\alpha}$ type spectrum ($\alpha \sim 2$) at regular intervals (periods) ranging from a fraction of a second to a few seconds. Subsequent work by other workers resulted in the discovery of more pulsars (Maran and Cameron, 1969).

The period lies between 1/30 sec to 3.7 seconds. In cases where accurate determination of the period can be made, it is found that the period P lengthens at a rate $dP/dt = 10^{-12}$ to 10^{-16}, and in no case does the period shorten regularly. The lengthening of the period shows that the time keeping mechanism is rotation and not oscillation (which is the cause for variability of some stars), for in the case of rotation the period lengthens as a result of dissipation of rotational energy and in the case of oscillation, a linearized theory will give rise to a constant period independent of the amplitude and when nonlinear effects are considered, the period decreases as the amplitude is damped, such as the case of a pendulum.

---

*Also to appear in the Proceedings of the IAU Symposium No. 46, The Crab Nebula, August 5-7, 1970 at Jodrell Bank.

If the basic time keeping mechanism is rotation then the value of the period gives an upper limit to the size of the pulsars. The centrifugal force per unit mass at the equator of a rotating star is $\omega^2 R$ and the gravitational force is $- GM/R^2$ when R is the radius, ω the rotational angular velocity, M the mass of the star. In order that the rotation be stable, it is necessary that $\omega^2 r < GM/R^2$, which gives R < 100 Km for the case of the Pulsar NP 0532 which has the shortest period of 1/30 sec.

Individual pulses are detectable in the Arecibo 1000 ft. radio telescope. It is found that the signal strength and structure changes from pulse to pulse, and in the case of NP 0532 microtime structure with a time constant as short as 100 μsec has been detected. If the dimension of the emitting region is R, then the time structure of the signal emitted cannot have a time constant shorter than the light transit time t_c which is R/c. For NP 0532, $t_c \sim 100$ μsec and this means that R < 30 Km (Drake and Craft, 1968).

No known astronomical objects other than asteriods are known to have a physical dimension less than 1000 Km (the white dwarf has a minimum theoretical dimension of 10^8 cm = 10^3 Km) except the neutron star, which was predicted by Landau in 1932 soon after the discovery of the neutron. At high density the Fermi energy of the electrons will be high enough so that the following relation will proceed favorably where p and n may be free or bound to a nucleus.

$$e^- + p \rightarrow n + \nu \quad (1)$$

The inverse reaction is forbidden because the available electron states are occupied. As (1) proceeds the elements become richer and richer in neutrons. At a density of 10^{12}g/cm^3 and above, few nuclei can exist and all but a few protons are converted into neutrons. A small fraction of matter remains in proton and electron state so that the inverse of (1) is prevented. Stable neutron stars have densities between 10^{14} to 10^{16}g/cm^3. The upper limit is dependent on the detailed structure of nuclear interaction, which is poorly known at high densities. The radius of a neutron star is of the order of 10 Km.

The neutron stars are results of stellar collapse which also gives rise to the supernova phenomenon. A pulsar (with the shortest period of 1/30 sec) is found at the center of the Crab nebula, a well known remnant of a supernova which flared up in 1054 A.D. Many theories predict that the star remnants of supernovae are neutron stars, hence, in all probability, pulsars are neutron stars. This pulsar NP 0532 is the youngest one on record and is the only one which has been found to emit not only in the radio frequency but also in the optical and x-ray spectrum ($10^{13} \rightarrow 10^{18}$ Hz). The division of energy flux in the radio, optical and x-ray regions are roughly: 10^{31} ergs/sec, 10^{34} ergs/sec, and 10^{37} ergs/sec respectively.

The neutron stars are themselves objects of immense interest. First, during the formation of a neutron star, a normal star suffers a change of radius from 10^{11} to 10^6 cm, representing a compression ratio of 10^5. The angular momentum conservation law $I\omega = MR^2\omega$ = constant requires that the period of rotation be increased as R^{-2}. The rotation period of a normal star such as the sun is 26 days, and after collapse a period of rotation as short as 1 msec may be achieved. (The centrifugal force at the equator gives a limiting rotation period of about 1 msec.) The rotational energy of a neutron star U_R, is roughly:

$$U_R \approx \frac{1}{2} I\omega^2 \approx \frac{1}{2} MR^2 \omega^2 \approx 7 \times 10^{46} (M/\odot) (R/10 \text{ Km})^2 P^{-2} \text{ergs} \quad (2)$$

where $\odot$ = solar mass = 2×10^{33}g.

Many stars possess magnetic fields of the order of 10^3 G or greater. During compression, the magnetic flux lines of a plasma is conserved and increases as R^{-2}. If the loss of flux lines is small during formation, the magnetic field of a neutron star may be as high as 10^{13}G.

Finally, another piece of observational evidence supports the rotational neutron star hypothesis of pulsars. The nuclei in the surface layer of a neutron star can form crystalline-like lattices (resulting in a smaller Coloumb energy) and the melting temperature of these crystalline structures is of the order of 10^9°K or more for densities between $10^7 \to 10^{13}$ g/cm^3. The temperature of neutron stars is usually quite low ($\ll 10^9$ °K), hence, the outer layer of neutron stars is made of crystalline solid. As the rotation of the star slows down, the equilibrium configuration is changed and strains are present in the crystalline layer. When the strain exceeds the stress limit of the crystalline layer, sudden resettlement (starquake) will take place. This will cause a sudden change in the moment of inertia of the star, resulting in a change of period as well as a change in the phase of the pulse called a glitch. Glitches have been observed in two cases: the case of the Vela pulsar and the case of the Crab pulsar. The glitch phenomenon, though easily explained in the rotating neutron star model, is virtually impossible to explain in oscillating models.

2. The Magnetic Field of the Neutron Star

A field as strong as 10^{12} G is hard to contemplate, but there is good evidence that such fields exist. As Pacini (1968) and Ostriker and Gunn (1969) suggested, a rotating neutron star with a magnetic field can emit electromagnetic radiation of frequency the same as that of the rotation. The energy of emission comes from the energy of rotation U_R. The rate of radiation dU/dt in the case of a rotating dipole is:

$$\frac{dU}{dt} = - \frac{2\bar{M} \sin^2 \theta \, \omega^4}{3c^2} \quad (3)$$

where $\bar{M}$ is the magnetic dipole moment, which can be approximated by

$$\bar{M} \approx \frac{\phi^2 R^2}{4\pi^2} = \frac{B_p{}^2 R^2}{4} \quad (4)$$

where ϕ is the flux line through the pole ($\phi \sim \beta_p R^2/\pi^2$ where β_p is the average field at the pole), θ is the angle of inclination of the dipole with respect to the rotational axis.

Equating the rate of loss of rotation of energy to dU/dt, it is found that

$$\frac{d\Omega}{dt} = - \frac{\phi^2 \sin^2 \theta \, R^2 \, \omega^3}{6\pi^2 \, I \, c^2} = 2\pi \frac{dP}{dt} \quad (5)$$

and

$$\Omega = \Omega_o \; (1+2t/\tau_o)^{-1/2}$$

For the Crab Nebula,

$$\omega \sim 100 \text{ sec}^{-1}, \; \theta \approx \pi/4, \; R \approx 10\text{Km}, \quad B_p \approx 10^{12}\text{G} \tag{6}$$

which agrees with observations.

The frequency of the emitted dipole radiation is the same as the pulsar. Such a radiation field is similar to the field in accelerators, and can accelerate particles to high energy. This explains the puzzling fact that the Crab nebula appears to be still active in producing high energy particles (which are needed to account for radio, optical, and x-ray emission from the nebula). Without the classical dipole radiation it will be very hard if not impossible to explain the rather large rate of loss of rotational energy of the pulsar. Analysis of other pulsars whose values of dP/dt are known all yield a field of the order of 10^{12} G. It thus seems well established that neutron stars may have fields as strong as 10^{12} G.

Goldreich and Julian (1969) pointed out that in the rest frame of the rotating magnetic neutron star there is a component of the electric field parallel to the magnetic field H such that the invariant $\underline{E}\cdot\underline{H} \neq 0$. The field strength at the surface is

$$E \sim H\frac{v}{c} \sim 10^{10} \; (H/10^{13} \text{ G}) \; (v_r/10 \text{ Km/sec}) \text{ volt/cm} \tag{7}$$

where v_r is the rotational velocity at the magnetic pole. This field will accelerate electrons and ions to stream out of the surface. As will be seen below, this electric field is also important to account for the radio (and optical and x-ray) emission from pulsars.

The properties of an electron gas in an intense magnetic field have been extensively studied (Chiu, 1970). First, the motion of electrons in a strong field is quantized in the direction perpendicular to the field and the usual expression of the kinetic energy of an electron is replaced by the equation

$$\varepsilon = mc^2 \; [1 + (p_z/mc)^2 + 2n(H/H_q)]^{1/2} - mc^2 \tag{8}$$

where the field is taken to be in the z direction, n is the quantum number characterizing the size of the classical orbit, $H_q = m^2c^3/e\hbar = 4.4 \times 10^{13}$G. Therefore, the momenta in the direction perpendicular to the field, $p_\perp$, is quantized by the relation:

$$p_\perp{}^2/m^2c^2 \rightarrow 2n(H/H_q) \tag{9}$$

In the nonrelativistic limit we have

$$\varepsilon = p_z{}^2/2m + n(H/H_q) \; mc^2 \tag{10}$$

The value of (H/H_q) mc^2 is 11.6 KeV if $H = 10^{12}$ G. If the electron kinetic energy is less than (H/Hq) mc^2, the only allowable state for the electron is the state n = 0, and then

$$\varepsilon = p_z^2/2m \tag{11}$$

which is the expression for the energy of a one-dimensional particle. The density of states for a one-dimensional electron is

$$\rho_1(\varepsilon) = \frac{2}{h}\frac{dp}{d\varepsilon} = \frac{2m}{h} p_z^{-1} \tag{12}$$

For an electron in the absence of field

$$\rho_3(\varepsilon) = \frac{2}{h^3}\frac{d^3p}{d\varepsilon} = \frac{2m}{h^3} p\int d\Omega \tag{13}$$

where $d\Omega$ is the solid angle element in the direction of the emerging particle. Therefore for a one-dimensional particle $\rho_1(\varepsilon) \to \infty$ as $p \to o$, and for a three-dimensional particle $\rho_3(\varepsilon) \to o$ as $p \to o$.

This one-dimensional behavior has been observed in low temperature experiments and this gives rise to the so-called deHass - Van Alphen effect. As a result of Eq. (9) there are only two radiation processes for "free" electrons in a magnetic field: (a) the spontaneous transition of an electron from a state n to n' $\neq$ n; (b) the Coulomb de-excitation of an electron from a state n to n' (no restriction on n' but the energy must be conserved). (a) is the quantized version of the classical synchrotron radiation. The emission gives rise to lines and the minimum energy of emission is $H/H_q\ mc^2$ (nonrelativistic case) or $(H/H_q)mc^2 \cdot mc^2/\varepsilon$ (relativistic case). The lifetime of electrons is generally very short; it is

$$\tau_s = \left[\frac{2}{3}\frac{e^2}{\hbar c}\frac{mc^2}{\hbar}\frac{\varepsilon}{mc^2}\left(\frac{H}{H_q}\right)^2\right]^{-1} \tag{14}$$

For $H \sim 10^{13}$ G, $\varepsilon \sim 1$ MeV, $\tau_s << 10^{-15}$ sec.

The bremsstrahlung radiation has been calculated by Goldman and Oster (1964), but their result is very difficult to apply. We have recently calculated the nonrelativistic bremsstrahlung radiation rate in a magnetic field (Canuto and Chiu, 1970). In this calculation the electron wave function in a magnetic field in conjunction with a Green's function appropriate for the field is used. This calculation is therefore expected to be valid for nonrelativistic electrons with energies up to, say, one or two hundred keV, in fields up to a fraction of $H_q = 4.414 \times 10^{13}$gauss, for frequencies not in the vicinity of the plasma frequency or the Larmor frequency of the electrons or ions.

The expression for the transition probability for the transition at forward angle is (Canuto and Chiu, 1970)

$$W = W_o\ c_1(\lambda)\ (pw)^{-1} \quad (cm^3\ sec^{-1}) \tag{15}$$

where $\varepsilon = E/mc^2$

$$W_o = Z^2\alpha^3\pi^2 N_i \lambda_c^3 \hbar c^2 (H/H_q)^{-2} = 2.09 \times 10^{-37}\ Z^2 N_i (H/H_q)^{-2} (erg\ cm^2\ sec^{-1})$$

$$\alpha = e^2/\hbar c, \quad \lambda\!\!\!^{-}_c = \hbar/mc,$$

N_i is the ion density and Z the average atomic number, and

$$C_1(\Lambda) = e^{\Lambda}(1+\Lambda)E(\Lambda) - 1, \; \Lambda = (\lambda\!\!\!^{-}_c{}^2/2\omega_H)\left[\left(\frac{\omega}{v}\right)^2 + d^{-2}\right] \tag{16}$$

$$E(\Lambda) = \int_{\Lambda}^{\infty} x^{-1} \exp(-x)\,dx \xrightarrow{\Lambda\to 0} \ell n\,(\gamma\Lambda)^{-1}$$

$$\gamma = 1.78102 \cdots \tag{17}$$

where ω_H is the Larmor frequency, v the electron velocity, d is the Debye length. This term d^2 arises from electron screening. (We assume the validity of classical electron screening theory). In the regime of interest $\Lambda << 1$ we can write

$$W = W_o(p\omega)^{-1}\ell n(\gamma\Lambda)^{-1} \tag{18}$$

The effect of a dense plasma is to alter the relation between the photon frequency ω and the wave vector $\underline{k}$ of the photon.

The dielectric constant is the same for a classical as well as a quantum electron gas, provided $\omega << \omega_H$ where ω_H is the Larmor frequency. The dielectric constant in a magnetic field is a tensional quantity. In the direction along the magnetic field the radiation can be split circularly into two circular polarization components, the ordinary component (denoted by a subscript o) and the extraordinary component (denoted by a subscript x):

$$n_x = 1 - \frac{\omega_p{}^2}{\omega^2}\,\frac{\omega}{\omega-\omega_H} - \frac{\omega_I{}^2}{\omega^2}\,\frac{\omega}{\omega+\Omega_H} \quad \text{(right hand polarization)}$$

$$n_o = 1 - \frac{\omega_p{}^2}{\omega^2}\,\frac{\omega}{\omega+\omega_H} - \frac{\omega_I{}^2}{\omega^2}\,\frac{\omega}{\omega-\Omega_H} \quad \text{(left hand polarization)} \tag{19}$$

ω_p and ω_I are the plasma frequencies of the electron gas and ion respectively, ω_H and Ω_H the Larmor frequency for electrons and ions. For $H = 10^{12}$ G, $\omega_H = 10^{19}$ and $\Omega_H = 10^{17}$, $\omega_p = 10^{16}\sqrt{\rho}$, $\omega_I \lesssim 10^{14}\sqrt{\rho}$, where ρ is the density. Hence, for $\rho \lesssim 10^6$ g/cm^3, n_x and n_o are close to unity for $\omega \sim 10^{10}$.

3. Radiation Mechanisms for Pulsars

From the upper limit of the radiating region of pulsars ($<3\times10^6$ cm) and the energy radio flux of emission (10^{31} ergs/sec) the flux of radiation I_ν at the emitting region is about 10^{18} ergs/cm^2 sec. The frequency of emission is at $\sim 10^3$ MHz. Assuming a bandwidth $\Delta f = f$, we find that the temperature T_B of a black body with an equivalent emissivity in the same frequency regime is given by:

$$I_\omega \sim 2kT_B \left(\frac{\omega}{2\pi}\right)^2 c^{-2} \tag{20}$$

or

$$T_B = \frac{I_\omega c^2}{2k(\omega/2\pi)^2}$$

Therefore, for pulsars the brightness temperature is about 10^{24} °K. If this radiation is emitted by a random processes then T_B is the minimum temperature of the emitter. The unduly large value of T_B requires that the process of emission cannot be an incoherent process.

Several laboratory devices can give rise to large brightness temperature in a relatively cool medium. These are lasers or Klystrons. In a Klystron an energetic electron beam bunched in space mingles with an electromagnetic wave with such a phase relationship that the electrons are decelerated by the electric field of the electromagnetic wave, and thereby the electrons give their kinetic energy to the electromagnetic wave. In this process the electrons must have such a spatial distribution that they are always decelerated, otherwise the electromagnetic wave will give energy to the electrons. For this reason, only electromagnetic waves of wavelengths within a bandwidth 1/2 λ are amplified.

This emission mechanism is also referred to as coherent plasma emission and has been considered as a possible emission mechanism for pulsars. There are a number of reasons why this mechanism may not work. First, the coherence of a plasma extends only over a spatial region of 1/2 λ; this also implies that the bandwidth of emission is of the order of 1/2 λ. Observation shows that the bandwidth extends from 50 - 1000 MHz, much beyond this bandwidth. Second, the intensity of radiation depends critically on the plasma density. Since the density of plasma surrounding the star depends on the structure of the magnetosphere which depends critically on the period P, it's then expected that the intensity of radiation should also depend on the period of radiation. However, observation shows that there is no such correlation. Third, observation shows that the optical and x-ray emission come from the same emission region. Therefore, the same plasma must be also responsible for optical and x-ray emission.

It is not possible to explain the (so far) unique case of optical and x-ray emission of NP0532 unless very artificial conditions are imposed. The other mechanism that we shall discuss here avoids all these difficulties. In the case of a laser, emission takes place between two atomic states whose energies are ε_1, ε_2 ($\varepsilon_1 < \varepsilon_2$). Coherence is achieved via a stimulated emission process, and the condition for coherent stimulated emission is such that the system has a negative temperature. Let the population of these two states be N_2 and N_1. Then an excitation temperature T_e may be defined by the equation

$$\frac{N_2}{N_1} = e^{-(\varepsilon_2 - \varepsilon_1)/kT_e} \tag{21}$$

The condition for $T_e < o$ is that $N_2 > N_1$. This case is referred to as population inversion.

The mass absorption coefficient $\kappa(\omega)$ with stimulated emission is given by

$$\kappa(\omega) = N_1 \frac{\sigma c}{\rho} (1-N_2/N_1) \tag{22}$$

where σ is the absorption cross-section, ρ is the density. $\kappa(\omega)$ therefore becomes negative when $N_2 > N_1$. The radiative transfer equation

$$\frac{dI}{ds} = - \kappa(\omega)\rho I \tag{23}$$

will have an exponentially growing solution

$$I_\omega = I_o \exp \int [-\kappa(\omega)]\rho \, ds \tag{24}$$

Radiation thus can be amplified. However, a laser operated between two atomic states again only gives rise to line emission. In order that a broad band emission be produced, it is necessary that the negative absorption coefficient extends over a finite frequency range. To see how this condition can be fulfilled, we can rewrite Eq. (21) as follows:

$$\frac{N_2 - N_1}{N_1} = e^{-(\varepsilon_2 - \varepsilon_1)/kT_e} - 1 \tag{25}$$

Let us consider a continuum and let $N_2 \rightarrow N_1$. Then we can write $N_2 - N_1 = dN$, $\varepsilon_2 - \varepsilon_1 = d\varepsilon$, and expand the exponential. We find

$$\frac{dN}{N} = - \frac{d\varepsilon}{kT_e} \tag{26}$$

or:

$$\frac{dN}{d\varepsilon} = - \frac{N}{kT_e} \tag{27}$$

The requirement for a negative temperature is then $dN/d\varepsilon > 0$. (Note: N refers to the occupation number per state and not the number of particles per unit energy range). However, since the condition $dN/d\varepsilon > 0$ cannot be maintained throughout the entire energy range (otherwise the total number of particles will necessarily be infinite) somewhere $dN/d\varepsilon$ will become less than zero. The condition for a negative absorption coefficient is more complicated than that for discrete states. The expression for the absorption coefficient $\kappa(\omega)$ for an electron gas in the lowest Landau level is

$$\kappa(\omega) = \frac{2m\hbar\omega}{\rho} \int_0^\infty gh^{-1} \, dp \left[f(p) \frac{\partial}{\partial p} \frac{\psi(p)}{p+p'} + f(-p) \frac{\partial}{\partial p} \frac{\psi(p)}{p-p'} \right] \tag{28}$$

where $\psi(p)$ is the transition probability for the absorption of a photon of energy ω by an electron of momentum p, p' is the electron momentum after absorption of a photon, f(p) is the electron distribution function, g is statistical weight factor of the electron (g is a function of the magnetic field) ρ is the density, $\psi(p)$ is related to W(p) [Eq. (15)] via the detailed balance theorem, which for a one-dimensional particle is simply

$$g\psi(p)\,dp = g\,W(p)\,dp', \quad dp'/dp = p/p' \tag{29}$$

Amplification of radiation (maser effect) is possible if $\kappa(\omega) < o$. The condition for this is not straightforward, as is clear from the form of equation (28). However, if the electron distribution function is asymmetrical, e.g., $f(p) >> f(-p)$ (this corresponds to a population inversion in terms of a coherent streaming motion in the +p direction with respect to the medium) then the condition for amplification of radiation is just:

$$\frac{\partial}{\partial p}\frac{\psi(p)}{p+p'} < o \tag{30}$$

or $\psi(p)$ must not rise faster than the first power of p. As is clear from the form of W for the $0 \rightarrow 0$ transition, equation (15), this condition is completely fulfilled. Hence, it is possible to have a negative emission coefficient.

4. A Simple Model

We choose f(p) a delta function at p_m [$f(p) = A\,\delta\,(p-p_m)$] where A is a normalization constant. ($A = g^{-1}\,N_e$ where N_e is the electron density.) We then have

$$\rho\kappa(\omega) = -\,2m\hbar c^{-1}\,N_e\,W_o\,p_m^{-3}\,\ln(\gamma\Lambda)^{-1} \tag{31}$$

The absorption coefficient is therefore negative and amplification of radiation takes place. The amplification is frequency independent, and covers the entire range of radio spectrum.

The radiation transfer equation is:

$$\frac{dI(\omega,\,s)}{ds} = -\,I(\omega,\,s)\rho\kappa(\omega) \tag{32}$$

whereas the spontaneous emission term is neglected. Eq. (32) then gives

$$I(\omega,\,s) = I(\omega,\,o)\,\exp\left\{\int_o^s p_m^{-3}\,\delta\,\ln[\gamma\Lambda]\right\} \tag{33}$$

$$\delta = 2m\hbar\,c^{-1}\,N_e\,W_o$$

and s is the path length over which amplification takes place.

Before we proceed further, we will discuss a very important effect in the theory of light and microwave amplification by stimulated emission. This is the effect of saturation of amplification. As it turns out, the amplification process is so efficient that the saturation phenomena takes place in all cases. The saturation phemomenon gives rise to non linear effects and in effect places the average brightness temperature within a range compatible with observations.

If a small signal is fed into an amplifier (such as an audio amplifier), the output power is proportional to the input signal. However, when the output power reaches the inherent power limit of the amplifier, further increase in the input signal level will not increase the output power. When this happens, the amplifier no longer amplifies the input signal linearly, and this amplifier is said to have reached saturation.

The same saturation effect also exists in amplification by stimulated emission.* Let us consider a two-level system. The reasoning can be easily extended to continuum energy level systems. Let there be n_2 particles in the upper energy state 2 and let n_1 be the number of particles in the lower state 1. In the absence of stimulated radiation, the lifetime of the state 2 is τ_2. Let $N = n_2 - n_1$ be the difference between particles in the upper and in the lower states, and population inversion is achieved when $N > 0$. Let N_o be the steady state value of N. (N_o depends on the pumping rate and other properties of the system.) The rate equation for N in the absence of stimulated radiation is:

$$\frac{dN}{dt} = \frac{N_o - N}{\tau_2} \qquad (34)$$

Now let us introduce stimulated radiation and let the transition probability via stimulated emission be P_{21}. The rate of change of N due to stimulated emission is simply $-2N\,P_{21}$. (The factor 2 arises from the fact that each transition from the state 2 to the state 1 changes the value of N by 2). The rate equation (34) now becomes

$$\frac{dN}{dt} = \frac{N_o - N}{\tau_2} - 2N\,P_{21} \qquad (35)$$

A steady state obtains when N_s satisfies

$$\frac{N_o - N_s}{\tau_2} - 2N_s\,P_{21} = 0 \qquad (36)$$

or when

$$\frac{N_s}{N_o} = [1+2\,P_{21}\,\tau_2]^{-1} \qquad (37)$$

As P_{21} is proportional to the intensity of stimulated radiation, $I(\omega, s)$, when $I(\omega, s)$ becomes large such that $2\,P_{21}\,\tau_2 \sim 1$, the value of N_s begins to decrease appreciably from the original value N_o and this laser amplifier no longer amplifies exponentially; the 'gain' decreases with increased output. Further increase in $I(\omega, s)$ will only cause a further decrease in the gain and, finally, as N_s approaches zero, the absorption coefficient being proportional to N, also becomes zero and the medium becomes

---

* G. Troup 1963, Masers and Lasers (London: Methuen)

transparent to radiation.

We can obtain the saturated intensity from the condition for the onset of saturation:

$$2P_{21}\tau_2 \approx 1 \tag{38}$$

P_{21} is the transition probability for emission per photon state and is easily seen from the Boltzmann equation for the photon distribution function to be just

$$P_{21} = F(\omega)\ \rho\ \kappa(\omega) \tag{39}$$

where $F(\omega)$ is the number of photons per state.

τ_2^{-1} is the transition probability for spontaneous emission, or collision, whichever is greater. The collision rate is about 10^{10} per second. The transition probability for spontaneous emission is much greater. The expression for τ_2^{-1} for spontaneous emission is

$$\tau_2^{-1} = N_e w(p) = 2.09 \times 10^{-37}\ Z^2 N_e N_i (pw)^{-1}\ (H/H_q)^{-2}\ C_1(\Lambda) \tag{40}$$

Numerically, if we set $N_e = 10^{24}$ cm^{-3}, $N_i = N_e/26$ (ion composition at the surface), $p = p_m = 0.01$ mc (see Section VIII), $H/H_q = 0.02$ ($H \sim 10^{12}$ gauss), $\omega \approx 10^9$ Hz, we find

$$\tau_2^{-1} \approx 10^{27} \tag{41}$$

The value of $\rho\kappa(\omega)$ which will give rise to saturation is about 40. Hence, from equation (39) we obtain the lower limit for the saturated value of $F(\omega)$

$$F(\omega) \sim 2 \times 10^{25} \tag{42}$$

The brightness temperature T_B is: [cf. Eq. (20)]

$$F(\omega) = \frac{kT_B}{\hbar\omega} \cong 10^2 T_B \quad (\omega \sim 10^9) \tag{43}$$

Comparing equation (42) with equation (43), we find that at saturation the lower limit of the brightness temperature is about 10^{23}°K. However, this value of T_B is an underestimate, for equation (39) gives the condition for the onset of saturation, and when saturation fully takes place, due to non-linear effects, the value of T_B may be several orders of magnitude greater than its value at the onset of saturation. The intensity of pulsar radiation is therefore independent on the period of pulsars, being a function of other parameters (such as the magnetic field, etc.).

5. Polarization of Radiation

In previous papers (see Chiu 1970) we have suggested that the propagation of radiation is strongly affected by the magnetic field resulting in a beamed emission. In this section we will discuss the dielectric properties of a plasma in a magnetic field.

As the frequency of radiation is small compared with the Larmor frequency of the ions and the electrons, as well as the plasma frequency of the electrons and the ions, we can use the dielectric constant computed for a cold ionic neutral plasma in a magnetic field. It turns out that quantum mechanical calculations and classical calculations coincide in this limit.

The dielectric tensor is (assuming that the magnetic field is in the z-direction)

$$\varepsilon_{\alpha\beta} = \begin{pmatrix} S & -iD & o \\ iD & S & o \\ o & o & \underline{p} \end{pmatrix} \tag{44}$$

where

$$R = 1 - \frac{\omega_p^2}{\omega^2}\frac{\omega}{\omega-\omega_1} - \frac{\omega_I^2}{\omega^2}\frac{\omega}{\omega+\Omega_H}$$

$$L = 1 - \frac{\omega_p^2}{\omega^2}\frac{\omega}{\omega+\omega_H} - \frac{\omega_I^2}{\omega^2}\frac{\omega}{\omega-\omega_H} \tag{45}$$

$$\underline{p} = 1 - \left(\frac{\omega_p}{\omega}\right)^2 - \left(\frac{\omega_I}{\omega}\right)^2, \quad 2D = R-L, \quad 2S = R+L \tag{46}$$

$$\omega_p^2 = 4ne^2N_e/m \qquad \omega_I^2 = 4ne^2ZN_i/m$$

$$\omega_H = eH/mc \qquad \Omega_H = Z\omega_H m/m_i \tag{47}$$

ω_p and ω_I are plasma frequencies of the electron gas and the ion gas, respectively, and ω_H and Ω_H the corresponding Larmor frequencies.

Generally speaking, the photon propagation in a doubly refractive medium such as equation (44) can be analyzed into two modes, the ordinary and the extraordinary mode of propagation. The propagation of these two modes can be easily studied for the direction $\theta = 0$ and $\theta = \pi/2$, respectively.

a). Along the magnetic field $\theta = 0$.

In this direction these two modes are right-handed (R) and left-handed (L) circularly polarized, and the refractive indices for these two modes are: (o = ordinary, x = extraordinary)

$$n_x^2 = R \tag{48}$$

$$n_o^2 = L \tag{49}$$

The emissivity of the x- and the o-modes in the 0 → 0 bremsstrahlung transition is:

$$\text{Ordinary mode:} \quad w^{(o)} = L^{1/2}\, w_o(p\omega)^{-1}\, C_1(\lambda)\, E_1^{-2} \tag{50}$$

$$\text{Extraordinary mode:} \quad w^{(x)} = R^{1/2}\, w_o(p\omega)^{-1}\, C_1(\lambda)\, E_2^{-2} \tag{51}$$

where

$$E_1^{-1} = (H/H_q)\left[\hbar\omega/mc^2 - H/H_q - \frac{1}{2}L(\hbar\omega/mc^2) - L^{1/2}(\hbar\omega/mc^2)\,(p/mc)\right]^{-1} \tag{52}$$

$$E_2^{-1} = (H/H_q)\left[\hbar\omega/mc^2 + H/H_q + \frac{1}{2}R(\hbar\omega/mc^2) - R^{1/2}(\hbar\omega/mc^2)\,(p/mc)\right]^{-1} \tag{53}$$

Now we will substitute into these equations quantities pertinent to a neutron star: $H = 10^{12}$ G, $N_e = 10^{24}$. We find $\omega_p \simeq 10^{16}$, $\omega_I \simeq 10^{14}$, $\omega_H \simeq 10^{19}$, $\Omega_H \simeq 10^{16}$. If we choose $\omega = 10^9$ Hz, we find that $E_1^{-2} = E_2^{-2} = 1$. For R and L we can expand equations (45) and (46) into power series in ω/ω_H and ω/Ω_H. For a charge neutral plasma, for which equation (47) is valid, we find that the first nonvanishing term in the expansion give:

$$R = 1 + \frac{\omega_p^2}{\omega_H^2} + \frac{\omega_p^2}{\Omega_H^2} \simeq 1 \tag{54}$$

$$L = 1 - \frac{\omega_p^2}{\omega_H^2} - \frac{\omega_p^2}{\Omega_H^2} \simeq 1 \tag{55}$$

and hence

$$w^{(o)}{}_{(p)} = w^{(x)}{}_{(p)} = w_o(p\omega)^{-1}\, C_1(\lambda) \tag{56}$$

Therefore these two modes have the same emissivity as that in vacuum. We therefore conclude that the radiation emerging from the dense plasma of the star along the magnetic field has both right- and left-hand polarization and has the same intensity. It is not known whether these two modes can combine into a linearly polarized beam or these two modes will propagate separately to yield an unpolarized beam.

b). Propagation perpendicular to the field, $\theta = \pi/2$.

In the perpendicular direction there is only one mode of propagation; this is the extraordinary mode. The refractive index for the ordinary mode is simply $1 < (\omega_I/\omega)^2 + (\omega_p/\omega)^2$, and propagation of this mode is forbidden. For the extraordinary mode the refractive index is:

$$n_x^2 = RL/S \approx 1 \tag{57}$$

and propagation is the same as in vacuum. This mode has one hundred percent linear polarization whose direction is perpendicular to both the wave vector $\underline{k}$ and magnetic field $\underline{H}$. There is another mode of emission whose electric field is along the magnetic field. Our computation shows that this mode is important in vacuum. However, since this mode cannot propagate in a dense plasma, it is not considered here.

c). Along an arbitrary direction.

As it turns out, the refractive index for the extraordinary mode is always close to unity for the field strength and plasma frequency we have quoted. The emerged radiation has an elliptical polarization which can be decomposed into a circular polarization plus a linear polarization. The polarization of the radiation from bremsstrahlung process that can emerge from the medium can either be circularly polarized, or linearly polarized, depending on the emissivity associated with these two polarization states. However, from simple arguments it seems that linear polarization is favored over circular polarization.

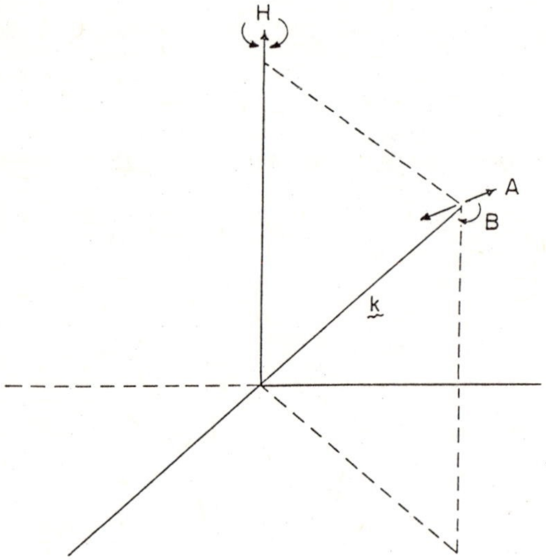

Fig. 1 - Polarization of the bremsstrahlung radiation in a strongly magnetized medium. k = propagation vector, H = magnetic field. A is the linearly polarized component perpendicular to the field; B is the circularly polarized component of the extraordinary component.

Figure 1 shows the directions of polarization of the emerging radiation. It is clear that if we look into the direction of the magnetic field, the radiation will appear unpolarized, assuming that these two modes (o and x) propagate and become amplified independently of each other. However, if we look at the medium at an oblique angle, the radiation will appear partially linearly polarized and the degree and inclination of polarization with respect to a fixed plane of rotation will depend on the relative orientation of the emitting surface with respect to the observer. This is illustrated in Figure 2. Thus, in general, we expect that pulsar radiation will be linearly polarized and the polarization strength and orientation will change through the cycle. Observation shows that in some cases pulsar radiation is found to be polarized and the orientation and the degree of polarization change with the phase of the pulse (Radhakrishnan and Cooke, 1969). We are now in the process of extending our computation of the emissivity of radiation at an arbitrary angle so that a comparison of our theory with observation can be made.

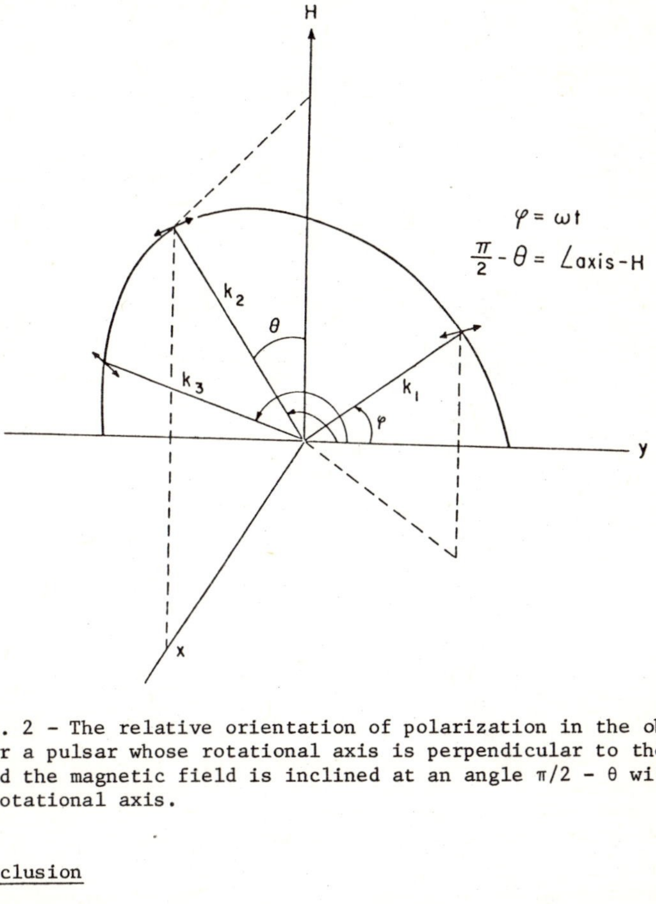

Fig. 2 - The relative orientation of polarization in the observer's plane for a pulsar whose rotational axis is perpendicular to the line of sight and the magnetic field is inclined at an angle $\pi/2 - \theta$ with respect to the rotational axis.

6. Conclusion

In this paper we have presented a model for pulsars. This model is based on the behavior of electrons in an intense magnetic field. While the model has not developed into a full quantitative treatment, it successfully arrived at a mechanism for emissions of intense radio emission. This mechanism avoids the usual difficulty of coherent plasma emission, that is the emission be limited to a bandwidth of $1/2\lambda$. A full account of this model is being published (Chiu and Canuto, 1970).

REFERENCES

Canuto, V., and Chiu, H. Y. 1970, Phys. Rev. D 188, 2446.
Chiu, H. Y. 1970, Publ. of Astron. Soc. of the Pacific 82, 487.
Chiu, H. Y., and Canuto, V. 1970, Ap. J. (to be published).
Drake, F. D., and Craft, H. D 1968, Nature 220, 231.
Goldman, P., and Oster, L. 1964, Phys. Rev. 136A, 602.
Goldreich, P., and Julian, W. H. 1969, Ap. J. 157, 869.
Hewish, A., Bell, S. J., Pilkington, J. D. H., Scott, P.F., and Collins, R. A. 1968, Nature 217, 709.
Maran, S. P., and Cameron, A. G. W. 1969, Earth and Extraterrestrial Sciences 1, 1.
Ostriker, J. P., and Gunn, J. 1969, Ap. J. 157, 1395.
Pacini, F. 1968, Nature 219, 145.
Radharkrishnan, V., and Cooke, D. J. 1969, Ap. Letters 3, 225.

November 18, 1969

Revised October 1, 1970

MAGNETOSPHERE THEORY OF PULSAR ELECTRODYNAMICS*

Peter Goldreich

California Institute of Technology
Pasadena, California

I want to begin by discussing the unipolar inductor, for example, an iron sphere, uniformly magnetized with magnetization parallel to the rotation axis. Let the sphere spin with angular velocity Ω and let brushes be placed at the pole and at the equator with a wire connecting the brushes. Because the sphere is an electrical conductor and its material is moving in a magnetic field (the fact that the field is the sphere's own field makes no difference) a magnetic force $q(\underline{\Omega} \times \underline{r}) \times \underline{B}/c$ is exerted on each free charge in the sphere. The charges therefore redistribute themselves within the sphere and on its surface in such a way that they set up an electric field interior to the sphere $(r < R)$ given by

$$\underline{E} = -(\underline{\Omega} \times \underline{r}) \times \underline{B}/c, \qquad r < R \tag{1}$$

The total electromagnetic force on each charge then vanishes. This electric field exists independent of whether the sphere is surrounded by conducting fluid or what have you.

Assume for the moment that we have a vacuum outside the sphere. The electric field outside the sphere is then given by a solution of Laplace's equation, matching potentials or tangential components of the electric field across the surface. The external potential takes the form

$$\Phi = -(B_o \, \Omega \, R^5/3cr^3) \, P_2 \, (\cos \theta), \qquad r > R \tag{2}$$

B_o is the field at the pole $(\theta = 0)$ and P_2 is the Legendre polynomial of second degree. In general there will be a discontinuity of the radial component of $\underline{E}$, implying a surface charge density

$$\sigma = -(B_o \, \Omega \, R/4\pi c) \cos^2 \theta, \qquad r = R \tag{3}$$

*Acknowledgment - The subject of this lecture is largely based on work I have done with Drs. D. Keeley and W. Julian.

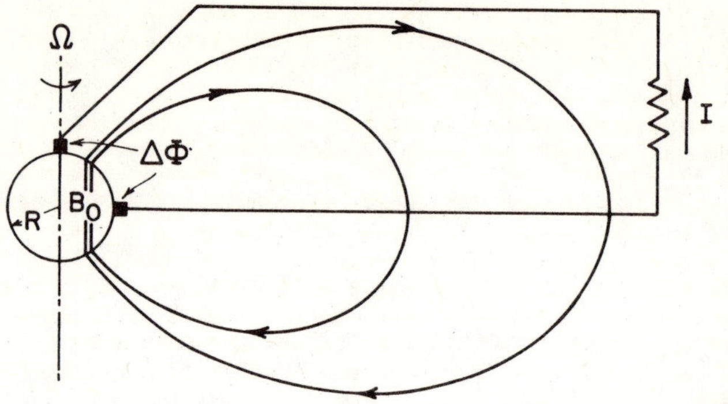

$r<R, \ \underset{\sim}{E} = -(\underset{\sim}{\Omega} \times \underset{\sim}{r}) \times \underset{\sim}{B}/c$

$r=R, \ \sigma = -(B_0 \Omega R/4\pi c)\cos^2\theta$

$r>R, \ \Phi = -(B_0 \Omega R^5/3c\, r^3)\, P_2(\cos\theta)$

$r<R, \ \underset{\sim}{E}\cdot\underset{\sim}{B} = 0$

$r>R, \ \underset{\sim}{E}\cdot\underset{\sim}{B} = -(\Omega R/c)(R/r)^7 B_0^2 \cos^3\theta$

	Iron magnet	Neutron star
R (cm	10	$10^6 R_6$
Ω (sec^{-1})	10^3	$2\pi/P$
B_0 (gauss)	10^4	$10^{12} B_{12}$
$\Delta\Phi$ (volts)	5	$3\times10^{16} B_{12} R_6^2/P$

Fig. 1. Unipolar inductor.

Incidentally, this expression for σ holds only if the magnetic field is continuous at the surface of the sphere. In the interior $\underline{E} \cdot \underline{B} = 0$ and outside the sphere

$$\underline{E} \cdot \underline{B} = -(\Omega R/c)(R/r)^7 B_o^2 \cos^3\theta, \qquad r > R \qquad (4)$$

if we assume a dipole magnetic field. This says that outside the sphere there is a component of electric field parallel to the magnetic field.

If we did this in the laboratory with values, say, $\Omega = 10^3$ sec^{-1}, $R = 10$ cm and $B_o = 10^4$ gauss then we would generate a potential difference $\Delta\Phi = 5$ volts between the pole and the equator. This would then drive a current through the circuit and dissipate energy in any resistance in the circuit, converting the mechanical energy of rotation into thermal energy. If we started the sphere rotating and didn't apply any further torque, the rotation would slow down. If you examine the current path in the sphere you will find that inside the sphere there is a $\underline{j} \times \underline{B}$ force that brakes the rotation. That's roughly how pulsars work.

If we scale this object up to neutron star dimensions we now have a potential difference given by

$$\Delta\Phi = 3 \times 10^{16}\ B_{12}\ R_6^{\ 2}/P \qquad (5)$$

volts where B_{12} is the field in units of 10^{12} gauss, R_6 is the radius in units of 10^6 cm and P is the period in seconds. So this object is obviously an interesting candidate for accelerating cosmic rays.

If we rotate a magnet in a vacuum in the laboratory nothing very unusual happens; the vacuum remains a vacuum. But this is not the case if we have a neutron star rotating in a vacuum. The surface charges implied by equation (3) could not be in equilibrium and would be drawn off the surface to form a magnetosphere. Assume there is a surface charge layer with a finite thickness. Magnetic field lines pass through the layer. Inside the star the Lorentz force vanishes and $\underline{E} \cdot \underline{B} = 0$. Outside the star $\underline{E} \cdot \underline{B} \neq 0$ if there is a vacuum. The component of $\underline{E}$ along the magnetic field lines is of order 3×10^{10} volts cm^{-1}, a respectable electric field. As we move along $\underline{B}$ from the interior outwards, the parallel electric field $\underline{E}_{\parallel}$ grows in a continuous fashion from zero to $\sim 10^{10}V\ cm^{-1}$ in a very short distance. There is no other force to compete with the electric force, which is always outward. The only force that could possibly hold the charges down is gravity, which is 10^{10} times weaker. The particles will therefore shoot out along the magnetic field lines.

The charges drawn off the surface will tend to distribute themselves so that the magnetic field lines become equipotentials. Then $\underline{E} \cdot \underline{B} \simeq 0$ outside as well as inside. If you think of this as an initial value problem with the sphere spun up from rest then at first negative charges would be shot out (if $\underline{\Omega} \cdot \underline{B} > 0$ inside the sphere). This would modify the external potential so that the surface charge layer would now contain positive charges which would also be shot out. Eventually you would end up with both sign charges outside in the right distribution to short out the field lines.

When this condition has been achieved we can distinguish two regions within the light cylinder. One region, called the magnetosphere, contains magnetic field lines that close. This is the region that corotates. The other region contains field lines that go out past the light cylinder. I can use the potential distribution on the surface, which is independent of what's outside the star, and the fact that the potential is continuous across the surface to figure out the potential everywhere else. Given the magnetic field structure and knowing that field lines are equipotentials I have the potential everywhere outside. From the potential I can find $\underline{E}$ and hence the space charge density. The number of electronic charges per cm^3 is given by

$$n = 7 \times 10^{-2}\ B_z/P \tag{6}$$

where B_z is in gauss. So a rotating neutron star will have a high density (by astrophysical standards) plasma around it.

We now consider the motions of these particles surrounding the star. If their energies are not too high, their motions are constrained to a sliding along the rigidly rotating field lines, like beads on wires fixed in the star. Any gyration around $\underline{B}$ is very quickly damped out by radiation. The motion of the particles is given by this simple picture just because $\underline{E} \cdot \underline{B}$ vanishes. The corotation of the magnetosphere probably has no observable consequences in pulsars. However, the rotation of this space charge contributes a toroidal current which is an important source for the magnetic field near the light cylinder. Therefore, the field will be significantly modified by these charges, even though their kinetic energy density is low compared to the magnetic field energy density.

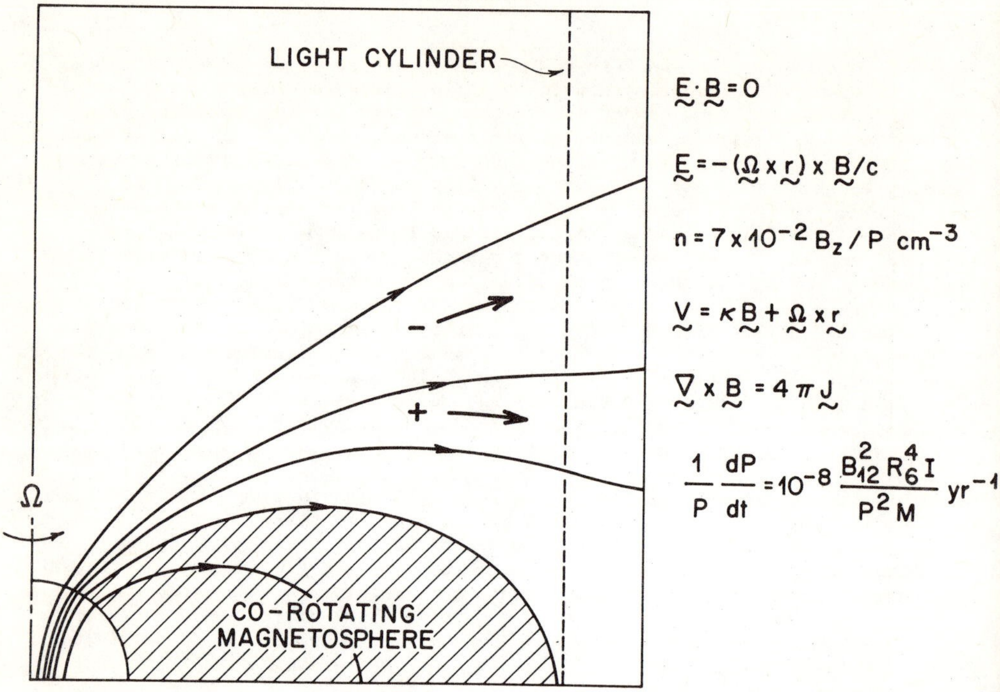

Fig. 2. Schematic diagram showing the corotating magnetosphere and the wind zone in which particles are accelerated and escape.

There is almost complete charge separation in the space charge region. To lift a proton about one stellar radius off the surface takes ~100 MeV but the gas at the surface cannot have a kinetic temperature anything like 100 MeV (if it did we would see hard quanta in emission). Therefore the plasma around the star should be thought of as cold. Cold plasma can only be held in place near the star by potential differences of order 10^8 V along the field lines, which is small compared to the 10^{16} V potential differences across the field lines. If you put a charge of the wrong sign into a region it would either fall down to the stellar surface or be flung out along its field line; it couldn't stay in rotation.

It's much more difficult to say what happens on the field lines that pass through the light cylinder. The plasma on these field lines cannot be in pure corotation because this would require particles that, beyond the light cylinder, moved faster than the speed of light and that's something that even pulsar astrophysicists don't want to tolerate. Nor can we have a corotating space charge on the segments of the open lines interior to the light cylinder and no space charge on the segments exterior, because that would require a surface charge layer on the light cylinder which would be unstable for exactly the same reason the layer on the surface of the star was unstable. But there has to be space charge in this region (for the same reason that there is space charge in the corotating region) and since it isn't corotating it must stream out.

Since electromagnetic forces dominate in this problem, at any instant only charges of a single sign will be accelerated along a given line of force. The charges receive their energy entirely from the rotation of the star, therefore there will be a torque on the star,

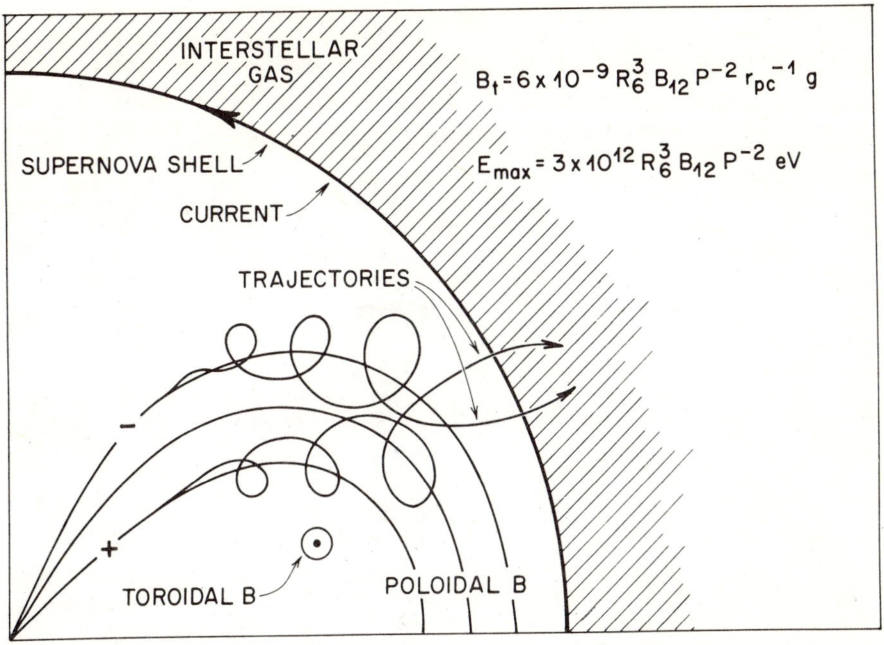

Fig. 3. Schematic diagram showing the supernova cavity, shell and interstellar gas.

slowing it down. (The torque slows the star down at a rate that gives the right order of magnitude for pulsar slow downs if you put in about 10^{12} gauss. Except for a slight difference in the coefficient it's the same deceleration rate as is given by magnetic dipole radiation).

If we assume a steady state, with equal numbers of charges of each sign flowing away, then it can be shown that because the poles are at lower potential than infinity, the negative charges flow out from the poles. Another way to see that the negative charges have to come from the polar regions is to calculate the magnetic field that is produced by the outflowing currents, which is a toroidal field. We find that if we assume that the negative charges flow out from the polar region while positive charges flow out from the lower latitude regions, then the Maxwell stresses on the surface of the star would decelerate the rotation. If we were to assume that the positive charges flowed out from the polar region and the negatives from the lower latitude region then the Maxwell stresses would accelrate the rotation, violating energy conservation. Therefore we know that the negative charges come out of the polar region. The trouble is, it turns out that if you really analyze it you can't have a true steady state. So by a steady state I mean some long-time average, say over a microsecond.

Now, what happens to these particles that escape? How much acceleration do they get and how do they radiate? The space charge is determined by the requirement that it make the lines of force equipotentials but along the open field lines the sign of the escaping charges is determined by the considerations mentioned above. It turns out that there is a region where positive charges are flowing out and the space charge has to be negative (if $\underline{E} \cdot \underline{B} = 0$), and vice-versa. This leads me to believe that one of two things must happen. One possibility is that there is a lot of neutral plasma around or opposite-sign-charge plasma around and the escaping charges are streaming through charges of opposite sign. I think that's sort of a bad idea, although at one time I didn't think it was such a bad idea. . .for a bad reason. The second possibility if that there isn't compensating space charge and these lines of force are not equipotentials. That is, in spite of the star's attempt to make these lines equipotentials all that happens is that particles are blasted out by the electric fields along B. This is the approach I want to take now. I want to calculate how energetic the particles will be if there is no space charge and they stream out along the magnetic field lines. That's the easy calculation. I'm going to do the one that I can do rather than the one that I should do, which is the self-consistent solution. . . using the continuity equation, calculating the particle distribution and using the Lorentz force and so on to calculate the acceleration. I can't do that because I'm not smart enough so I'm going to make an assumption.

I look at the region of the polar cap where there are field lines that go through the light cylinder. This cap has an angular size given approximately by

$$\theta \simeq (\Omega R/c)^{1/2} \qquad (7)$$

which is typically about 0.01 to 0.1 radians. This may have something to do with the duty cycle of the pulsar, that is, with the duration of the pulse. I calculate the order of magnitude of the acceleration under the assumption that the corotating region is all shorted out, i.e., $\underline{E} \cdot \underline{B} = 0$. Then there has to be a potential difference across the cap given by

$$\Delta\Phi \simeq \frac{1}{2} B_o R \left(\frac{\Omega R}{c}\right)^2 \qquad (8)$$

Given this potential difference, with "conducting walls" represented by the highest latitude corotating field line, what is the electric field in the region bounded by the polar cap, the rotation axis and the "conducting walls?" The solution of this problem is

$$\left|\frac{\underline{E} \cdot \underline{B}}{B}\right| = B_o \left(\frac{\Omega R}{c}\right)^{3/2} \exp\left[-\frac{(r-R)}{R}\left(\frac{c}{\Omega R}\right)^{1/2}\right] \qquad (9)$$

This field falls off very quickly with height and is negligible when you reach a height of about one cap radius.

If we put one electron into this field then it goes out along the magnetic field and gets accelerated. For a pulsar that has a period $P = 0.5$ sec, a field 10^{12} gauss and a radius 10 Km, the electron is accelerated to an energy $\varepsilon = \gamma mc^2$ with $\gamma \simeq 10^8$. For fast pulsars like the Crab γ may reach about 10^9 but it can't get very much higher because radiation damping starts to become important. As the electron moves along the curved magnetic field line it tends to radiate. The electrons will emit γ-rays having a characteristic synchrotron spectrum with a maximum near the critical frequency ω_c, with $\hbar\omega_c/mc^2 \simeq 10^5$.

These are very hard photons ($\sim 10^5$ MeV) and we don't see them. That's because the photons produce pairs. The process by which they do this is well understood although it is not too simple to calculate. You don't have to worry about the guys who did it though--one of them was the former chairman of this department, John Toll, in his Ph.D. thesis. What happens is that these hard photons are emitted essentially tangential to $\underline{B}$ but because the lines are curved they soon see a component of $\underline{B}$ perpendicular to their momentum. Then they can produce pairs. The rate of pair production depends on exp $-(8/3\chi)$ where

$$\chi = \frac{\hbar\omega}{mc^2}\frac{B_o}{B_{cr}} \tag{10}$$

and

$$B_{cr} = \frac{m^2c^3}{e\hbar} = 4 \times 10^{13} \text{ gauss.} \tag{11}$$

Thus, as soon as the photon sees a substantial B_o the rate becomes very large. The pairs are therefore produced very close to where the photon is generated. These then radiate more photons which make more pairs and so on.

At one time I had a very clever idea which I can't resist telling you about even though it's wrong. What happens to the pair is that the electric field drives the electron out and drives the positron inward. Now you have the original electron plus the new electron, separated by a few centimeters and moving together. (Everything is going at the speed of light.) They produce more γ-rays, which produce more electrons, the positrons always draining back toward the star. This is a way of building up big charge bunches to make coherent radiation. But it's really not a good way because all the positrons going backward emit γ-rays, which produce more pairs and the electrons from the secondary showers start going out and the average electron isn't in a very big shower.

For moderate pulsars this breakdown of the vacuum can only be initiated by electrons. Protons can never get up to high enough energies to produce hard enough photons to produce pairs. For fast pulsars like the Crab however, the protons can reach sufficiently high energies and produce this breakdown of the vacuum. For such pulsars most of the outflowing positive charge may be positrons not protons. So you might think that in the Crab nebula there are no protons produced.

The charge separation that is set up by this process sets up an electric field that opposes the original field. So the process shuts off, then the pair production shuts off and we start over again. So that's a proof that we can't have a steady state, unless we don't reach these high energies.

I want to end by showing how the radio emission might be made. The radio emission from individual particles streaming out this way is completely negligible. We also know that the radiation has a brightness temperature reaching 10^{28} °K and if you have an incoherent radiation process you need particles with energies about 10^{24} eV and even for me that's too much; I can't get such high energies. So we have to think about some coherent radiation processes. Now in order to produce the meter-wave radiation that we see we need a bunch containing about 10^{25}

electrons radiating coherently. A bunch like this spread over 1 meter parallel to B and 50 m perpendicular to B (the radius of the cap) would develop an enormous internal electric field of about 10^{10} V cm^{-1} and would be violently unstable. However, if we're pushing this bunch with an external field of about 10^{11} V cm^{-1} then it doesn't matter if the bunch produces its own field because the external field is stronger. In a fraction of a centimeter the bunch reaches the velocity of light. Since the particles can only go at the velocity of light, they will all stay very close to each other even if the electric field strength is different on the two ends of the bunch. So big electric fields are conducive to the stability of charge bunches because the particles are pushed up against the velocity of light and they don't blow apart. How charge bunches get formed in the first place is more difficult to understand. If you try to satisfy the dynamical equations and the continuity equation for a charge layer which is exploding off the star you will see that the particles come off in a bunch that is essentially as thin as the bunch can be and still be stable. So it's quite suggestive that large electric fields produce bunches that can then produce intense radio emission.

One final remark. I wondered a long time about the gamma radiation that would be produced by incoherent radiation and isn't seen. The very high energy γ-rays wouldn't get out because they produce pairs but the soft ones, below 100 MeV, could get out. At least for the Crab there is a problem of why we don't see these γ-rays. This is very serious so I dreamed up a process for photon splitting, which turned out to have actually been calculated. . .but incorrectly. A photon with energy of order, say, 1 MeV moving in a magnetic field can produce a virtual pair which might emit a synchrotron photon before coalescing. Skobov (1959) calculated a cross section for photon splitting and if you take his calculation seriously then it predicts that even those photons that aren't hard enough to produce real pairs could still split into softer ones. In the case of the Crab, γ-rays would be degraded all the way into the optical range if they were emitted near the surface.

The trouble with Skobov's calculation (aside from some numerical mistakes and the omission of some photon polarization vectors) is that it is not gauge invariant. The correct perturbation theory answer (at least to lowest order in B) is zero. However the amplitude for photon splitting almost certainly doesn't vanish if you treat the magnetic field not as a perturbation but instead use the exact wave functions for an electron in a magnetic field. Well, I'm going to try to do that although it looks pretty hard. So I don't know about photon splitting but some of you may be interested in trying to play with it.

DISCUSSION

L. BAGGERLY: "Well, what makes your pulsars pulse?"

P. GOLDREICH: "I knew somebody would ask that. Everything I said about the acceleration in the near field is independent of whether the magnetic axis is tipped or aligned with the rotation axis. If the particles are accelerated near the pole and the radiation comes out in a pencil, then if the magnetic axis is tipped the pencil swings around and if you're lucky it hits you."

REFERENCE

Skobov, V.G. 1959, Soviet Physics JETP, 8, 919.

December 2, 1969

PULSARS AND THE ORIGIN OF COSMIC RAYS

J. P. Ostriker

Princeton University Observatory

The work I will discuss has been done in collaboration with James E. Gunn, also of Princeton. I think you will discover as we go along that most of the general energetics are common to all the quantitative theories. That is Pacini, Goldreich and Julian and ourselves make the same predictions for the ages, magnetic fields, and energy output of the pulsars. There are, however, differences in details. As I would like to fit the pulsar phenomenon into the general astronomical picture, you may find me depart somewhat from the topic promised by my title.

As you know, the energy spectrum of cosmic rays covers an enormous range of energy, with a rough power law. At the high energies the differential spectrum is about $dn \alpha E^{-2.5}dE$. The maximum energy observed seems to be about 10^{20} to $10^{20.5}$eV.

Regarding the Crab nebula I want to stress several points. We know when it exploded and its expansion rate. Most of the luminosity is due to continuous emission, snychrotron radiation, by electrons with energies on the order of 10^{13}eV moving in fields on the order of 10^{-3} to 10^{-4}G. The filaments are $\sim$ 1-2 $M_\odot$ of ordinary hot plasma. Embedded in the nebula is a star that's blinking on and off, the pulsar (NP 0532). The pulsar contributes only a small fraction ($\sim 10^{-2}$) of the total luminosity of the nebula, even less if you include the expansion energy in the total. I stress the Crab nebula because it's the best observed object. We know something about the particles coming out of it, its age and so on.

However, we also know about as much about pulsars as a <u>class</u> as we know about the Crab as an individual object. There are about 40 of them. Their periods, P, range from 33 ms to about 4 s. Their characteristic ages, defined as $P/\dot{P}$, range from $\sim 10^3$ years to $\sim 10^8$ y. As you all know, with one or two temporary anomalies, the sign of $\dot{P}$ is always the same, positive. Certainly over long periods of time $\dot{P}$ is positive. The radio luminosities of the pulsars are very low, typically $L_{radio} \simeq 10^{-4} L_\odot$. They are not bright objects. Finally, two of them are in supernova remnants.

Now Wheeler suggested in 1966, that is, in the year 2 B.P. (Before Pulsars), that there was a rotating neutron star in the Crab. So, if you will, what we've all been doing is spelling out his suggestion. Now Wheeler may or may not have been led to his suggestion this way, but it is possible to argue that if the energy output from the Crab is $\sim 10^6 L_\odot$ as it is and the source of the energy is a rotating star then the star just <u>has</u> to be a rotating neutron star. The argument is rather simple. If one considers a log M - log R (M = mass, R = radius) plot, then there are several restraints on where the object in the Crab can lie on this plot. First of all it must be bigger than its own Schwarzchild radius. That is

$$R > \frac{2GM}{c^2} \tag{1}$$

That is, M/R must be less than a certain number. Secondly, assuming it rotates, one requires that it be in equilibrium and not be subject to fission. That is

$$\Omega^2 < 0.229\ G\ \bar{\rho} \tag{2}$$

where $\bar{\rho}$ is the mean density and Ω is the angular velocity. Now we know Ω, therefore M/R^3 must exceed a certain value. Finally, if we require that the energy of the whole nebula be supplied by the loss of rotational energy of the object then

$$\dot{E} = I\Omega\dot{\Omega} \tag{3}$$

where I is the moment of inertia. Now we know $\dot{E}$, Ω and $\dot{\Omega}$ so that we know something about the moment of inertia. Then taking a range of values for the ratio of the radius to the radius of gyration puts upper and lower limits on MR^2. These four conditions limit the object to a small region on the log M - log R diagram in which only a neutron star can exist.

These roughest of arguments thus tell you that it has to have the mass, density and radius characteristic of neutron stars. Actually this had been thought of a long time ago. Baade and Zwicky (1934) suggested it; Oppenheimer and Volkoff (1939) calculated it; Colgate, Hoyle, Cameron suggested that neutron stars would be left after supernova explosions. So it was an extraordinary situation in astrophysics. Normally the theoretical "predictions" follow the observations. This was the once-in-a-lifetime, or several life-times, where the scientific method worked as it's supposed to, as in the textbooks. On the basis of theory people predicted that there should be a certain kind of object and lo-and-behold, thirty years later it was found.

Well, what should the characteristics of neutron stars be? Different people make different models. Kip Thorne for example has calculated the radius of cold matter as a function of the total mass. There are two stable regions, where $dM/dR < 0$, corresponding to white dwarfs ($R \simeq 10^3$ to 10^4 km) and to neutron stars ($R \simeq 10$ km). If you ask what mass the neutron star is likely to have the answer is it should have a mass greater than the maximum mass of a white dwarf so that in the collapse it doesn't stop at the white dwarf stage and less than $\sim 1.8\ M_\odot$. This puts the mass on the order of 1.2 to 1.4 $M_\odot$ and $R \simeq 10$ km.

The neutron star is also likely to have a strong magnetic field. Main sequence stars have magnetic fields and if the material is a sufficiently good conductor then the magnetic flux is frozen into the matter and BR^2 = constant as it collapses. Therefore in going from a solar radius to 10 km, a 100-Gauss field becomes $\sim 10^{12}$ G. Incidentally, one can ask if such a large field would dramatically affect the star? Would it perhaps blow it apart? Well, you can show by very simple scaling arguments that the magnetic field energy scales the same way as the gravitational energy, i.e., as R^{-1}, so that if the star didn't mind its magnetic energy when it was on the main sequence it won't mind it when it's a neutron star. Therefore, a neutron star with $B \sim 10^{12}$ G will be very nearly spherical.

Now, as Pacini (1968) and others (including us) pointed out, a spinning dipole in vacuum loses energy by magnetic dipole radiation at the rate

$$\dot{E}_{md} = \frac{2m_\perp^2\Omega^4}{3c^3} \qquad (4)$$

where $m_\perp$ is magnetic-dipole moment perpendicular to the rotation axis. Also, if the star is distorted and has a dumbbell component to its shape there will also be gravitational quadrupole radiation at the rate

$$\dot{E}_{gq} = -\frac{GD_\perp^2\Omega^6}{45c^5} \qquad (5)$$

where $D_\perp$ is the component of the mass-quadrupole moment perpendicular to the rotation axis. Finally, any object losing energy by multipole radiation also loses angular momentum at the rate

$$\dot{J} = \dot{E}/\Omega \qquad (6)$$

therefore it must slow down.

We don't know $m_\perp$ or $D_\perp$ but we can estimate $m_\perp$ to be of order $R^3B \simeq 10^{30}$esu. If we just take the $\dot{E}_{md}$ term and (6) we can integrate the equation for $\dot{P}$ and get

$$P(t) = P_o(1 + 2t/\tau_o)^{1/2} \qquad (7)$$

where $\tau_o = (P/\dot{P})_o$ is a characteristic time evaluated when $P = P_o$. Notice that we don't need to know the moment of inertia to get (7).

We can now apply this to the Crab and integrate back in time. We find that about 1100 years ago it should have had a very short period. But we know that the age of the Crab is only 915 years. There are then two alternatives at this point. We can say the theory is quite good but let's not worry about errors of 20 percent, or, we can try to patch it up. If you want to patch it up you must imagine another loss mechanism that was more powerful in the past than it is now, something that depends on a higher power of Ω. The gravitational radiation does just that. So you can put in a small amount of gravitational radiation, which won't make very much difference now but which will make the Crab extrapolate back to the right starting point.

In order to choose between the two models (with or without gravitational radiation) you need to know $\ddot{P}$. The magnetic dipole radiation model predicts $\ddot{P} = -0.016$ ns day^{-1} yr^{-1} if gravitational radiation is unimportant, or $\ddot{P} = -0.021$ ns day^{-1}yr^{-1} if it is important. Well, that's something one can check.

Furthermore, if we believe we know the moment of inertia and we observe Ω and $\dot{\Omega}$, then we can get everything else. For example we can compute the magnetic dipole luminosity $L_{md} = 8 \times 10^{38}$ erg s^{-1}. We can also get the magnetic field. We can tell how the luminosity varied in the past. From (4) and (7) we find that the luminosity varies as $L_{md} \propto t^{-2}$. Therefore the Crab was much, much, brighter in the past.

How much magnetic dipole radiation will it put out over its lifetime? The answer depends on what you say about the gravitational radiation. If the gravitational radiation is unimportant then essentially all the kinetic energy of the neutron star ends up coming out as magnetic dipole radiation and we get $\Delta E_{md} \simeq 7 \times 10^{52}$ ergs. If you put in the gravitational radiation of a sufficient amount to make the theoretical and observed ages agree, then you get $\Delta E_{md} \simeq 5 \times 10^{51}$ ergs. Assuming an initial $\Omega \simeq 10^4$ then the initial rates of magnetic-dipole and gravitational-quadrupole radiation are $L_{om} \simeq 5 \times 10^{45}$ ergs s^{-1} and $L_{og} \simeq 3 \times 10^{48}$ ergs s^{-1}. These rates are comparable to supernova luminosities. We can also find the field strength by assuming a field such that it gives the magnetic-dipole rate that slows down the star at the observed rate. This gives $B = 2.5 \times 10^{12}$ gauss.

If we're going to believe this theory it must do more than explain the Crab. You could say: if the Crab has this magnetic field and I suppose other pulsars have similar fields, are their rates of slowing down consistent with this B? Or you can turn the question around and say: given the theory and the conservative assumption that they all have the same moment of inertia then what fields are required to explain the observed values of $\dot{P}$? If this "coupling constant" turns out to be extraordinarily variable then you just won't believe the theory. Well, we find that even though the periods range over two orders of magnitude, the derived fields only vary by a factor two or so, clustering around 10^{12} G (Fig. 1a). When we take the decay of the magnetic field into account, which I will come to later, it turns out that the original distribution of field strengths was even tighter (Fig. 1b) than what we see now.

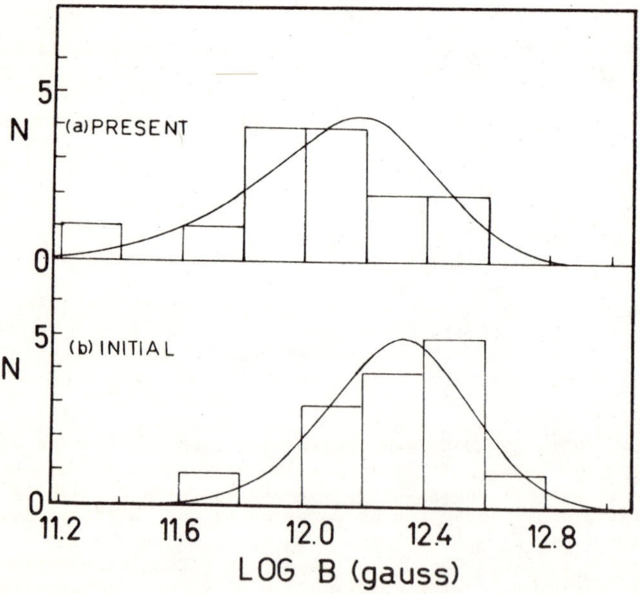

Fig. 1. Distributions of derived present (upper) and initial (lower) surface magnetic field strengths. The solid lines are the predictions of theory as described in the text.

The picture then is very similar to what Pacini and Goldreich-Julian are saying. A large fraction of the kinetic energy of a rotating neutron star may be expected to be emitted in this low frequency electromagnetic radiation. This radiation is below the plasma frequency so it's not going to propagate into interstellar space. It may push on the nebula and expand it but it can't get out. However, it may do other things. Goldreich and Julian (1969) have argued very convincingly that the neutron star must emit particles. What happens to these particles? Gunn and I have computed the magnetic field around the rotating neutron star assuming it is a dipole field close to the star and we then integrated the relativistic equations of motion of a test particle in a vacuum. We start our particle off in the wave zone, far from the star.

The physics of the particle acceleration involved is very simple. Since the gyrofrequency $\omega_0 = eB/mc$ is very much larger than the wave frequency Ω, particles injected in the wave zone find themselves in a strong, nearly static, crossed electric and magnetic field. In a time very short compared to the period of the wave they are moving relativistically in nearly the direction of the wave's propagation vector. Thereafter they ride the wave at essentially constant phase. Actually, that's a homework problem in "Landau and Liwschitz;" it's not a new problem. If you put in the coriolis forces and the centrifugal forces it becomes much more complicated but the solution is actually simpler; the energy of the particle turns out to be independent of where you put it in, provided it's far into the wave zone to begin with. The particle gets an energy E given by

$$E = 3.6 \times 10^{13} A(Z/A)^{2/3} (\Omega_2)^{4/3} (B_{s,12})^{2/3} (R_6)^2 \quad \text{eV} \qquad (8)$$

where Z, A are the atomic number and atomic mass of the particle, Ω_2 is the neutron star's angular velocity in units of 10^2 s^{-1}, $B_{s,12}$ is the surface field in units of 10^{12} G, and R_6 is the star's radius in units of 10^6 cm. For the Crab this gives energies of $\sim 2 \times 10^{13}$ eV for electrons and $\sim 2 \times 10^{14}$ eV for protons now. Well, these are happy numbers because the synchrotron spectrum implies electrons of about 10^{13} eV in the Crab.

The Goldreich-Julian theory, translating the numbers slightly, gives a rate of injection of particles

$$n_i = \frac{B_s \Omega^2 R^3}{Zec} = 7 \times 10^{32} (\Omega_2)^2 (B_s, _{12}) (R_6)^3 Z^{-1} \qquad (9)$$

particles sec^{-1}.

Given the injection rate and the particle energy in terms of the angular velocity, and the angular velocity (or the period) as a function of time (eqn. [7]), we can now apply these ideas to the Crab nebula and make certain predictions. These can be compared to the observations. This is done in Table 1.

This theory predicts that the Crab is now producing electrons with energy $\sim 10^{13}$eV, protons with energy $\sim 1.6 \times 10^{15}$eV and heavy ions like Fe with energy $\sim 5.5 \times 10^{16}$eV. Early in its history the proton energies were $\sim 3 \times 10^{17}$eV and Fe-ion energies $\sim 10^{19}$eV. Therefore the Crab has been producing high-energy cosmic rays throughout its lifetime. The energy spectrum of these cosmic rays can be computed from the history of the object. Assume a delta-function injection at the energy given by equation (8) and write the instantaneous injection rate, the particle energy and the rate of change of angular velocity as powers of the angular velocity

$$n_i = A\Omega^{\alpha}, \qquad E = B\Omega^{\beta}, \qquad \dot{\Omega} = C\Omega^{\gamma} \tag{10}$$

Then relate the particles emitted in a given time interval to the energy interval spanned during that time

$$N(E)\,dE = n_i(t)\,dt = n_i(t)\,\frac{dE}{dt}^{-1}\,dE \tag{11}$$

where $dE/dt = (dE/d\Omega)\dot{\Omega}$. From (10) and (11) we find that the energy spectrum is

$$N(E)\,dE = D\,E^{\delta}\,dE \tag{12}$$

where

$$\delta = (\alpha + 1 - \beta - \gamma)/\beta \tag{13}$$

From the theory of Goldreich-Julian we have $\alpha = 2$. Our theory gives $\beta = 4/3$ and, early in the history when the high-energy particles are produced, $\gamma = 5$. Then $\delta = -2.5$, compared to the observed spectral index for high energy cosmic rays of -2.6.

TABLE 1

PROPERTIES OF THE CRAB NEBULA AS PREDICTED BY THIS THEORY AND AS OBSERVED

	Predicted by Theory	Observed	Remarks
Age	1180 y	915 y	
Electron synchrotron luminosity	7×10^{37} erg s^{-1}	1.2×10^{38} erg s^{-1}	
Total energy	8×10^{38} erg	$(4\text{-}12)\times10^{38}$ erg	Mostly in kinetic energy of expansion
Field at edge of nebula	10^{-3} G	$(1\text{-}10)\times10^{-4}$ G	
$-\ddot{P}$	.016-0.21 ns/d/y	.009 ns/d/y .040 ns/d/y	Arecibo Princeton

The number density of cosmic rays depends on the rate of pulsar production. The pulsar birth rate seems to be about one per 30 years in the Galaxy, if it's a 15 kpc uniform disk. If these particles are injected into the appropriate metagalactic volume for the Hubble time you get about 100 times more intelgalactic cosmic rays than are observed. However, our galaxy may be rather overenergetic in supernovae. Within the uncertainty of the numbers, there's no difficulty in producing the requisite number of particles and the spectrum looks right.

Finally, I point out that the theory presented here predicts an upper limit to the energy of the particles produced. This is obtained by maximizing everything possible in equation (8). The fastest possible rotation is obtained by setting

$$\frac{1}{2} I\Omega^2 = GM/2R$$

setting magnetic energy equal to gravitational energy, and setting R = the Schwarzchild radius. Then we find

$$\frac{E_{max}}{mc^2} \approx \left(\frac{e^2}{G\ m^2}\right)^{1/3} \simeq 10^{12}, \tag{14}$$

a famous cosmological, mysterious constant, implying $E_{max} \simeq 10^{21}$ eV. Therefore, if we find particles with energies considerably above this value the theory is in trouble.

To summarize the cosmic ray problem it seems quite promising to produce the high-energy cosmic rays by this process, in terms of the spectrum, the number and their maximum energy. As far as the predicted composition goes we have almost no information. However, this mechanism cannot work for the <u>bulk</u> of the cosmic rays; by the time the pulsar is rotating slowly enough to make low-energy cosmic rays it doesn't have enough energy left. You either have to have another mechanism to produce lower energy particles or you have to degrade the energy of the higher energy ones.

Now let's ask, what happens to the pulsar? Because the electrical conductivity is finite, the magnetic field will decay with time. The dipole moment will decay according to

$$\text{dipole moment} \propto e^{-t/t_d},$$

where $t_d = 4R^2\sigma/\pi c^2$. Because the field decays, the radiation shuts off and the period stops increasing. The period obeys the law

$$P(t) = P_o \left[1 + \frac{t_d}{t_o}(1 - e^{-2t/t_d})\right]^{1/2} \tag{15}$$

The conductivity is the subject of some controversy; that of Canuto (1970) seems to be okay for the outer layers of the star. His value is $\sim 10^{23}\text{sec}^{-1}$, giving a decay time $t_d \approx 4 \times 10^6$ yrs. The best estimate for σ in the mantle may be that given by Solinger (1970) and his value is very close to Canuto's. The interior condutivity is much higher, but, after the currents in the outer parts of the star have decayed the surface field will be so much reduced that the magnetic dipole luminosity will essentially cease.

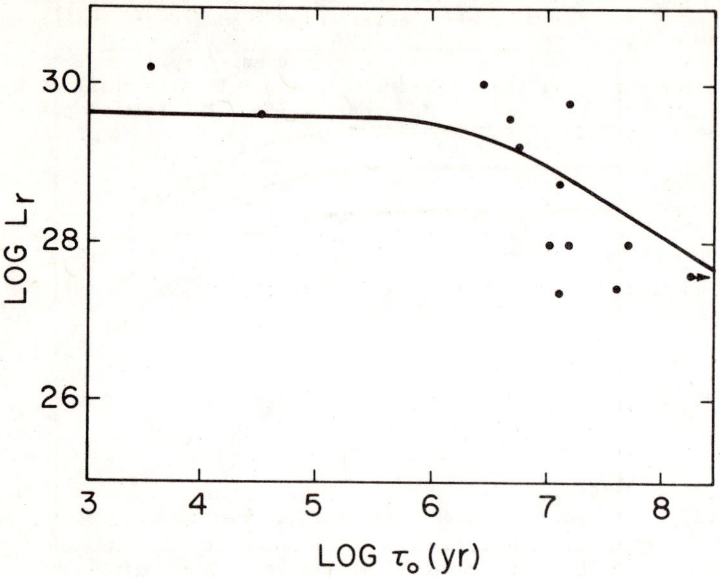

Fig. 2. Radio luminosity vs the characteristic time $\tau_o = P/\dot{P}$. The solid line is an "evolutionary track" for $t_d = 4 \times 10^6$ yr.

There is evidence that the radio luminosity decays, older pulsars tending to be weaker (Fig. 2). There is also a tendency for older (i.e., slower) pulsars to have weaker magnetic fields (Fig. 3). This is weak, direct evidence for magnetic field decay.

The indirect evidence for field decay is very strong. If the field didn't decay we would see many long-period pulsars. Well, there are 13 pulsars with periods between 1 and 2 seconds but only 1 pulsar with period longer than 2 seconds. Something has happened to shut off the pulsars after a few million years. The period distribution predicted by this theory fits very well to the observed distribution. The typical age is $\sim 2 \times 10^6$ years.

What kind of stars do pulsars come from? The distribution in distance from the plane of the Galaxy, derived from dispersion measures, shows an exponential distribution with scale height $\sim$ 160 pc. But if the pulsars are divided into four groups according to their periods, long-period pulsars are found to be furthest away from the plane. The best fit to the observed distribution is obtained if the initial scale height is 80 pc and the pulsars are born with an initial velocity, or kick, of 100 km sec^{-1} (Fig. 4). There are two observations that make such an initial kick rather probable. First of all, the Crab nebula pulsar has a proper motion corresponding to a tangential velocity $\sim$ 100 km sec^{-1}. Secondly, there is a category of stars, called "runaway" stars, discussed by Blaauw (1969), that have velocities of 100 - 200 km sec^{-1}. Blaauw explained such runaways as being the product of a binary system in which one of the members becomes a supernova, leaving its partner to run off. Well, if this happens and produces a neutron star then the neutron star will follow the runaway and have a velocity $\sim$ 100 km sec^{-1}. This is interesting for the radio observers because they should look for proper motions. They should be on the order of 0.1 second of arc per year for the closer pulsars. That would give us the possibility of determining statistical parallaxes of pulsars.

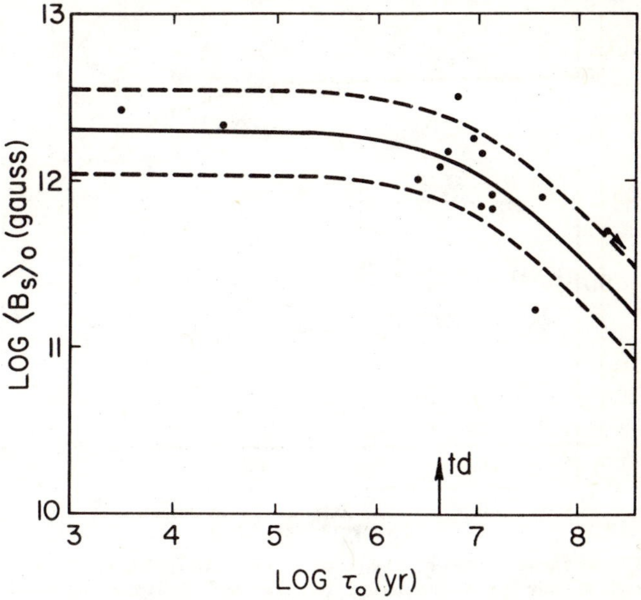

Fig. 3. Surface fields (deduced from observed P) versus the characteristic time $\tau_o = P/\dot{P}$. The solid line is an evolutionary track for $t_d = 4 \times 10^6$ yrs, $B_i = 2 \times 10^{12}$ G; the dotted lines are for initial fields of exp $(\pm\sigma_i)$ times this.

Since the pulsars are close to the plane they must belong to a young population - population I in the astronomer's jargon.

From the distances and ages one gets the rate of formation, 5×10^{-11} pulsars $year^{-1}$ pc^{-2} in the plane. This is 1 per 30 y in the Galaxy. This is very close to the supernova rate and to the rate of runaway stars in the Galaxy. It also agrees with Shklovsky's and Oort's figures derived from numbers of supernova remnants.

From the birth rate and the scale height we can estimate that the mass of the parent star must be between 2 and 10 $M_\odot$. If you say that all stars of mass greater than x become pulsars, then using the luminosity function and the birth rate function for the neighborhood of the sun, x is 4 $M_\odot$. Now, there are 7×10^9 low-mass stars in the center of the Galaxy. Let's suppose that when those stars were born they were in the same ratio to high-mass stars as in the solar neighborhood. Then this predicts a certain number of high-mass stars originally in the center of the Galaxy. If all the high-mass stars heavier than 4 $M_\odot$ became supernovae then the supernova rate was $\sim$ 0.1 to 1 per year for the early epoch of the Galaxy. This rate is very high and it would have remained high until the main sequence burned down to about 4 solar masses, which is about 10^8 years. The energy output in *non-thermal* form from supernova remnants would then have been somewhere between that of the Seyfert galaxies and the quasars,

$\sim 10^{44}$ ergs sec^{-1}. Thus our galaxy, extrapolated back in time, would have looked like a small source, showing fluctuations on a time scale of one year, and with a spectrum looking like that of the Crab, and a total luminosity like the less bright QSO's.

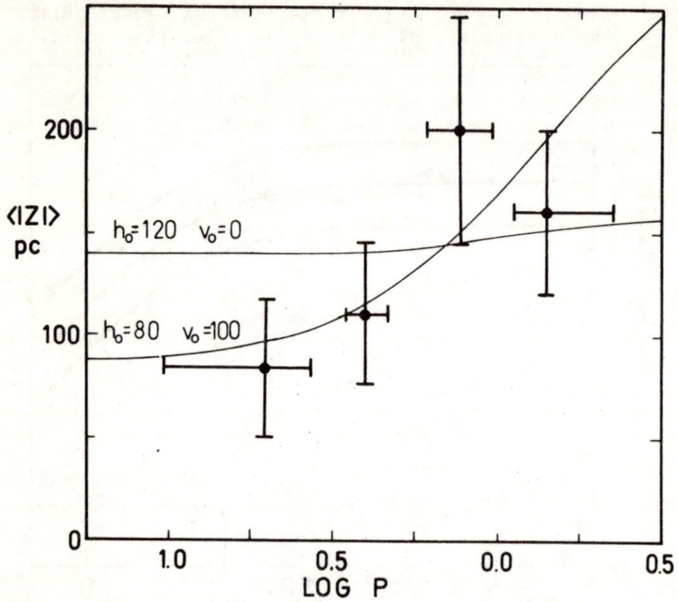

Fig. 4. The observed scale height as a function of period. The boundaries of the bins are log P < -0.5, 0.5 < log P < -0.25, 0.25 < log P < 0, and 0 < log P. The bars are 2 standard deviations high, the width is the approximate r.m.s. width of the objects in the bins, and the centers are placed horizontally at the mean log P of the objects in the bin. The nearly horizontal line is the expected mean height if h_o is 120 pc and there is no velocity kick at birth. The sharply rising line is for h_o = 80 pc and a mean kick of 100 km/sec at birth.

It is now tempting to soup the model up and ask what would the situation be like early in the history of giant elliptical galaxy - on which note I will end.

DISCUSSION

H. ZAPOLSKY: "You get into a little bit of trouble with your cosmic ray spectrum if you use $\gamma = 3$ instead of 5. Have you calculated the realistic value of perpendicular component of quadrupole moment to see whether or not at say, the break-up value of Ω, the gravitational radiation would still swamp the electromagnetic radiation?"

J.P. OSTRIKER: "First of all, in my capacity as a rotating star man, let me say there isn't any break-up. Now, a minimum value of the quadrupole moment would be that produced by the field we see. If there's no toroidal field and only the poloidal dipole-like field we see, then the energy loss will still be dominated by gravitational radiation initially. If the field is large enough to cause a significant pressure anisotropy then gravitational radiation dominates for most of the lifetime. Actually, the whole issue is a little bit of an embarrasment. It's too good. You start out with the rapidly rotating neutron star having $\sim 10^{53}$ ergs. Therefore, in anyone's theory you need a large energy sink since the observed output of supernovae is only $10^{51} - 10^{52}$ ergs. Gravitational radiation, early on, seems to be the most natural one I can think of."

F. STECKER: "The Cornell-Sydney group has done an analysis of the arrival directions of cosmic rays above 10^{18} eV and they are rather random."

J.P. OSTRIKER: "You wouldn't expect them to be anisotropic. When the pulsar is very young it produced very high energy particles. But there haven't been any supernovae recently. Therefore we shouldn't expect any high-energy cosmic rays from sources in the Galaxy now. Furthermore, if some other galaxies are like our galaxy, then given the inverse-square law the nearer sources have a weak effect compared to distant sources so that either way the component due to our galaxy is rather negligible."

M.M. SHAPIRO: "Can the bulk of the cosmic rays be produced by white dwarfs?"

J.P. OSTRIKER: "It's attractive in a lot of ways. The white dwarfs have a chemical abundance in their atmospheres that's much more reasonable for making cosmic rays than either neutron stars or supernovae. Also, if white dwarfs are born rapidly rotating as you might expect, they have plenty of energy. Also this mechanism scales perfectly well for the white dwarfs. So it's a distinct possibility. The only objection is that, in order to produce the power law that you observe, you would have to arrange the <u>initial</u> angular velocities of the white dwarfs in the right power law, because they don't change very much. However, if it turns out that you don't need to put in a power law injection to get out a power law spectrum then I'd be happy to bring the question up again. But it almost seems that you have to put in a power law to end up with one. Otherwise I think it's quite all right."

LESLIE FISHBONE: "Do you have a representative number for the particles injected from the pulsar?"

J.P. OSTRIKER: "Well, I have some numbers but I don't feel at all confident about them. They are taken, in a crude adaptation, from Goldreich and Julian's theory [cf equation (9)]."

J.R. WAYLAND: "As the cosmic rays come out through the expanding nebula they are degraded in energy and a delta-function injection becomes a power law."

J.P. OSTRIKER: "Would that it were so. For the electrons that's definitely the case, because of their synchrotron radiation. However, for the protons, which don't scatter much at these high energies, it may not be so."

T. STECHER: "Has there been any attempt to associate known pulsars with runaway stars?"

J.P. OSTRIKER: "Yes, we're trying to do it. But we only see a small fraction of the pulsars and it's difficult to do."

REFERENCES

Baade, W., and Zwicky, F. 1934, Proc. Nat. Acad. Sci., 20, 254.
Blaauw, A. 1961, B.A.N., 15, 265.
Canuto, V. 1970, Ap. J., 159, 641.
Goldreich, P., and Julian, W. 1969, Ap. J., 157, 869.
Oppenheimer, J.R., and Volkoff, G. 1939, Phys. Rev., 55, 374.
Pacini, F. 1968, Nature, 219, 145.
Solinger, A. 1970, Ap. J., 161, 553.
Wheeler, J. 1966, Ann. Rev. Astr. and Ap., 4, 423.

November 24, 1969

APPENDIX

APPROXIMATE PARAMETERS OF 55 PULSARS

S.B. Modali*

S.P. Maran

Laboratory for Solar Physics
Goddard Space Flight Center

The names, periods, equatorial coordinates (epoch 1950) and dispersion measures are given in the accompanying table for the 55 pulsars that have been announced as of September 1970. The available data on pulse widths are so imhomogeneous that we have not felt them worthy of presentation in this book. From time to time, lists of pulsars with these and other parameters given to the precision of the available observations are published in Earth and Extraterrestrial Sciences.

*Research Assistant at University of Maryland. Ordinarily at University of Texas at Austin.

TABLE 1

APPROXIMATE PARAMETERS OF 55 PULSARS

Source	Period (sec)	R.A. h	m	Dec °	'	$\int n_e ds$ cm^{-3} pc
MP 0031	0.943	00	31.6	-07	37	11
MP 0254	0.448	02	54.4	-54		10
PSR 0329+54	0.715	03	29.2	+54	25	27
MP 0450	0.550	04	50.4	-18	00	25
NP 0527	3.745	05	25.8	+21	58	51
NP 0532	0.033	05	31.5	+21	59	57
PSR 0628-28	1.244	06	28.9	-28	33	34
MP 0736	0.375	07	36.8	-40	34	100
CP 0808	1.292	08	09.0	+74	38	6
MP 0818	1.237	08	18.1	-15		25
AP 0823+26	0.531	08	23.8	+26	47	19
PSR 0833-45	0.089	08	33.7	-45	00	63
PSR 0834+06	1.274	08	34.4	+06	21	13
MP 0835	0.765	08	35.6	-40		120
PSR 0904+77	1.579	09	04	+77	40	-
MP 0940	0.662	09	40.7	-56		145
PP 0943	1.098	09	43.3	+10	05	15
PSR 0950+08	0.253	09	50.5	+08	10	3
MP 0959	1.437	09	59.8	-54	37	90
PSR 1133+16	1.188	11	33.5	+16	07	5
MP 1154	0.400	11	54.8	-62		270
AP 1237+25	1.382	12	37.3	+25	12	9
MP 1240	0.388	12	40.4	-63	36	220
MP 1359	0.690	13	59.7	-50		20
MP 1426	0.788	14	26.6	-67		60
MP 1449	0.180	14	49.4	-65		90
PSR 1451-68	0.263	14	51.5	-68	32	9
PSR 1508+55	0.740	15	08.1	+55	43	20

TABLE 1

APPROXIMATE PARAMETERS OF 55 PULSARS (continued)

Source	Period (sec)	R.A. h	m	Dec °	'	$\int n_e ds$ $cm^{-3}pc$
MP 1530	1.369	15	30.4	-53	30	20
AP 1541+09	0.748	15	41.2	+09	38	35
MP 1604	0.421	16	04.6	-03		10
PSR 1642-03	0.388	16	42.2	-03	13	36
MP 1706	0.653	17	06.6	-16	37	10
MP 1727	0.830	17	27.8	-47	40	121
MP 1747	0.742	17	47.9	-46	56	40
PSR 1749-28	0.563	17	49.8	-28	06	51
MP 1818	0.598	18	18.2	-04	28	70
JP 1845	0.598	18	45.2	-04	00	141.9
MP 1857	0.612	18	57.7	-25		35
JP 1858	0.655	18	58.7	+03	24	402
MP 1911	0.826	19	11.2	-04	47	75
PSR 1919+21	1.337	19	19.6	+21	47	12
PSR 1929+10	0.227	19	29.9	+10	53	8
PSR 1933+16	0.359	19	33.5	+16	10	143
MP 1944	0.440	19	44.6	+17		35
JP 1946	0.717	19	46	+35	30	100
JP 1953	0.427	19	53	+29	12	20
JP 2003	2.111	20	03	+31	30	225
PSR 2016+28	0.558	20	16.0	+28	31	14
JP 2022	0.529	20	22.3	+51	44	10
PSR 2045-16	1.962	20	45.8	-16	27	11
JP 2113	1.015	21	13.5	+46	36	100
PSR 2218+47	0.538	22	18.3	+47	40	44
AP 2303+30	1.576	23	03.5	+30	45	46
JP 2319	2.256	23	19.25	+60	00	140